D0058167

DARWIN FAMILIES

wood
5

```
h Elizabeth   Josiah II (Uncle Jos)  =  Elizabeth (Bessy) Allen
78-1856       1769-1843                 1764-1846
                                       ─ Sarah Elizabeth (Elizabeth, Aunt Sarah) 1793-1880
                                       ─ Mary Anne 1796-8
                                       ─ Charlotte    =  Charles Langton
                                          1797-1862      1801-86
                                                          └─ Edmund 1841-75
                                       ─ Henry Allen  =  Jessie Wedgwood
Frances Mosley  =  Francis              (Harry)          1804-72
   1808-74          1800-88             1799-1885      ─ Louisa Frances 1834-1903
Godfrey 1833-1905 ─                                    ─ Caroline b. 1836
Amy 1835-1910 ─                                        ─ John Darwin 1840-70
cely Mary 1837-1917 ─                                  ─ Anne Jane 1841-77
Clement 1840-89 ─                                      ─ Arthur 1843-1900
Lawrence 1844-1913 ─                                   └─ Rowland 1847-1921
nce Rose 1846-1903 ─                   ─ Hensleigh   =  Frances (Fanny) Mackintosh
l Frances 1852-1930 ─                    1803-91         1800-89
                                       ─ Frances      ─ Frances Julia (Snow)
                                         (Fanny)          1833-1913
                                         1806-32       ─ James Mackintosh (Bro)
                                                          1834-64
                                                       ─ Ernest Hensleigh 1838-98
                                                       ─ Katherine Euphemia (Effie)
                                                          1839-1931
                                                       ─ Alfred Allen 1842-92
                                                       └─ Hope Elizabeth 1844-1934
```

Charles Darwin stands as a towering figure in the history of science, who changed the direction of modern thought by establishing the basis of evolutionary biology. These letters offer a fascinating window into his daily experience, scientific observations, personal concerns and friendships, affording a unique glimpse of Darwin as both naturalist and family man. From his early years at Edinburgh University up to the publication of the *Origin of species* in 1859, the letters in this volume chart the most exciting years of Darwin's life, including the voyage of the *Beagle* and the subsequent findings which led to his theory of natural selection.

The Cambridge edition of the *Correspondence of Charles Darwin* (winner of the first Morton N. Cohen Award for a Distinguished Edition of Letters) has been hailed as a 'monumental, beautiful edition' and 'one of the triumphs, in scope and excellence, of post-war publishing'. This selection, compiled by the senior editor of the *Correspondence*, and introduced by Stephen Jay Gould, makes these engaging letters newly available to general readers.

Portrait of Charles Darwin, copy of a watercolour by George Richmond, 1840, Darwin Museum, Down House.

CHARLES DARWIN'S LETTERS

A SELECTION 1825–1859

EDITED BY
FREDERICK BURKHARDT

CAMBRIDGE
UNIVERSITY PRESS

Published by the Press Syndicate of the University of Cambridge
The Pitt Building, Trumpington Street, Cambridge CB2 1RP
40 West 20th Street, New York, NY 10011-4211, USA
10 Stamford Road, Oakleigh, Melbourne 3166, Australia

© Cambridge University Press 1996

First published 1996

Printed in Great Britain by Biddles Ltd., Guildford & King's Lynn

A catalogue record for this book is available from the British Library

Library of Congress cataloguing in publication data available

ISBN 0 521 56212 0 hardback

Contents

v

Contents

Illustrations

Foreword

A Life's Epistolary Drama

Museum of Comparative Zoology
Harvard University

This book should be viewed as an intense drama of a particular kind, and in an honored genre. Epistolary books do not appeal to everyone (though they rank among my personal favourites), but stories told as letters form an ancient and popular tradition from St Paul onward. Epistolary non-fiction embodies the added virtue of an actual life, as told (however truly, and whatever hidden) by the protagonist.

Many historians have lamented the decline of letters written on preservable paper as a frightful loss to the trade. Modern technology has placed us in a paradoxical situation. We usually assume that information becomes richer as we approach the present day — for more kinds of records ought to become available, and less deterioration and loss should occur. But we now send most of our communiqués but once over a telephone wire into permanent oblivion, and many (if not most) of our letters now exist only in the cyberspace of e-mail and other similar devices (where they may, of course, be transferred to paper, but will just as likely be dumped into a lost realm with our phone calls).

Charles Darwin's life (1809–1882) falls squarely into an intermediary time of maximal information — after an older time of too much information lost (and no adequate postal service), and before our modern age of electronic obliteration. The intellectual leaders of Darwin's day wrote voluminous letters; in optimal cases, we may have daily records of a scientific collaboration (see Martin Rudwick's *The great Devonian controversy*, for such a fascinating record of Sedgwick and Murchison's daily contact in their early collaborations on establishing a geological time scale — a rich tapestry that can rarely be equalled in any other time, even by oral historians working with living subjects today).

We must not be fooled into thinking that letters either speak accurately, or tell everything. In the continuum of hidden personal feelings and motives, letters occupy an intermediate status in a progressively more adequate sequence: published works, letters, and private journals and diaries. (Even the most private diaries need not attain the probably chimerical ideal of genuine accuracy. In defending himself before a Senate inquiry in 1993, a Washington political staffer actually testified that he had consciously lied to his own diary!) In Darwin's case, for example, we crucially miss any epistolary insight into his greatest intellectual eureka, the formulation of natural selection in 1838 — for this he kept truly hidden from others, and therefore recorded only in his private jottings.

Charles Darwin may not have been the greatest of Victorian prose stylists; these honors, I think, must go to Charles Lyell and Thomas Henry Huxley. But no one could possibly match Darwin for fascination and appeal — a remarkably genial man who must inspire affection, but also a strikingly complex and often cryptic person, whose contradictory desires and influences cannot be resolved into coherence, even by access to an epistolary domain more intimate than his published works. Above all, of course, Charles Darwin changed our intellectual world perhaps more than any other person in the history of science — irrevocably and (with respect to ancient psychological hopes and social traditions) most painfully, for all the joy of his great insight. For this cardinal reason alone, the wonderfully expressive and richly varied letters of Charles Darwin represent one of the great dramas of western history.

Many literary devices exist for organising an epistolary book, and I rejoice that Burkhardt has chosen the time-honored scheme of strict chronology — for temporal direction provides the most useful and principled theme of contingent histories (where key items do not occur as predictable consequences of nature's laws, but as largely fortuitous outcomes of undetermined antecedent strings of events), and human lives are contingent histories par excellence. (This volume stops in Darwin's midlife, with the publication of *Origin of species* in 1859. I surely hope that we shall soon have a second volume, for Darwin never sank into intellectual quietude, and the less overtly dramatic years of his old age are just as full of insight, and just as engaging, as his youthful adventure on the *Beagle*. The strength and

Foreword

poignancy of 'the faith that looks through death in years that bring the philosophic mind' can match all the mobility and eventfulness of global circumnavigation!')

If, continuing with Wordsworth (one of Darwin's favorites), 'the child is father to the man', then the events of a life, molding the flexibilities of inborn temperament, form the defining features of adult greatness — and chronology, again, becomes our best device for insight. (We do, after all, study the lives of eminent people, in part as pure voyeurs to be sure, but largely to touch the sources and mystery of such preeminent accomplishment.) An epistolary chronology may be optimal for understanding the dichotomous (and elusive) themes that construct a life: vagaries of development and the nature of stable personality.

On the theme of development, we can best trace Darwin's passage from awkward enthusiasm to confident manhood through the ideal (and extended) baptism of a five year voyage around the world on HMS *Beagle*.

As an unenthusiastic medical student in Edinburgh, just shy of his seventeenth birthday, he writes to his sister Susan about a great opportunity for gaining new knowledge in his real passion of natural history (and at little expense pain upon the pursestrings of his wealthy, though long-suffering, father): 'I am going to learn to stuff birds, from a blackamoor . . . it has the recommendation of cheapness . . . as he only charges one guinea, for an hour every day for two months' (p. 4). (Darwin's study with John Edmonstone, a freed slave, represents his first known contact with blacks, and this good experience helped to set his liberal views on race.) At age 19, now studying in Cambridge, he writes to his cousin W. D. Fox on the loneliness of an ardent collector: 'I am dying by inches, from not having any body to talk to about insects' (p. 6).

He sails on the *Beagle* at age 22, and quickly discovers the perils of ocean travel. Darwin writes to his father: 'the misery I endured from sea-sickness is far far beyond what I ever guessed at . . . I must especially except your receipt of raisins, which is the only food that the stomach will bear' (p. 16).

But the beauty and riotous diversity of the tropics extinguishes this memory of pain and Darwin positively bursts with joy: 'it would be as profitable to explain to a blind man colors, as to a person, who has not been out of Europe, the total dissimilarity of a Tropical view . . . So you must excuse raptures & those raptures badly expressed' (p. 17).

xi **PORTLAND COMMUNITY COLLEGE LEARNING RESOURCE CENTER**

Foreword

His old plans quickly fade (a restrictive and conventional scenario largely set by others). Already he realises that he shall probably not become a simple country reverend with an amateur's local concern for natural history: 'by the fates, at this pace I have no chance for the parsonage' (to his sister Caroline in April 1832). But growth also entails pain, as the love of his young life marries another (and rather quickly) in his absence: 'if Fanny was not perhaps at this time Mrs Biddulp, I would say poor dear Fanny till I fell to sleep.— I feel much inclined to philosophize but I am at a loss what to think or say; whilst really melting with tenderness I cry my dearest Fanny why I demand . . .' (p. 22).

Early in the voyage, Darwin decries his lack of experience: 'It is positively distressing, to walk in the glorious forest, amidst such treasures, & feel they are all thrown away upon one' (p. 23). But he collects with zeal and sends barrels of specimens back to his mentor J. S. Henslow in Cambridge: 'I am afraid you will groan or rather the floor of the Lecture room will, when the casks arrive' (p. 26).

As the voyage progresses, he gains confidence and even begins to see his inexperience as a blessing in disguise: 'By the way,' he writes to Henslow in 1834, 'I have not one clear idea about cleavage, stratification, lines of upheaval.— I have no books, which tell me much & what they do I cannot apply to what I see. In consequence I draw my own conclusions, & most gloriously ridiculous ones they are' (p. 31). He gains confidence in collecting: 'I have just got scent of some fossil bones of a **Mammoth**!, what they may be, I do not know, but if gold or galloping will get them, they shall be mine' (p. 34).

By the end of the voyage, Darwin has gained full confidence in his geological abilities. He writes to his sister Susan in April 1835: 'there is a strong presumption (in my own mind conviction) that the enormous mass of mountains . . . are so very modern as to be contemporaneous with the plains of Patagonia . . . If this result shall be considered as proved it is a very important fact in the theory of the formation of the world' (p. 43).

As the *Beagle* sails for home, Darwin finally visits a coral reef and validates a mechanism for the formation of atolls — his first important work in theoretical science — that he had been developing and revising for months: 'The subject of Coral formation has for the last half year, been a point of particular interest to me. I hope to be able to put some of the facts in a more simple & connected point of view, than that in which they have hitherto been considered' (p. 50).

Darwin returns to England at age 27, a full professional with the baptism of experience. He distributes his specimens to taxonomic experts for their description, and he completes the cycle of his own growth by advising his old mentor Henslow to publish one big paper, rather than several fragments. The teacher has become the taught: 'Do think once again of making one paper on the Flora of these islands [the Galápagos] . . . if your descriptions are frittered in different journals, the general character of the Flora never will be known, & foreigners, at least, will not be able to refer to this & that journal for the different species' (p. 69).

On the second theme of the formed personality and intellect that anchored Darwin's brilliance and success, I think that we can best capture the complexities (and occasional contradictions) of his persona by stating that Darwin was radical in his scientific ideas, liberal in his political and social views, and conservative in personal lifestyle — all within the context of a class background that mattered so much, and made so much possible in Victorian Britain: Darwin was a man of considerable independent wealth, and he came from an upper class country family of impeccable social standing.

Darwin's intellectual radicalism emerges most clearly in the nature of natural selection as a materialistic theory about a history of life without sensible purpose or necessary progress — not merely in the espousal of evolutionism in some form. In a famous letter containing one of his earliest 'confessions', Darwin writes to Hooker in 1844: 'I am almost convinced . . . that species are not (it is like confessing a murder) immutable. Heaven forfend me from Lamarck nonsense of a "tendency to progression" . . . but the conclusions I am led to are not widely different from his—though the means of change are wholly so' (p. 81).

As publication of the *Origin* approached, and now from the greater comfort and success of his middle age, Darwin still delighted in his power to shock, as in this playful comment to Asa Gray in 1857: 'To give one example, the last time I saw my dear old friend Falconer, he attacked me most vigorously, but quite kindly, & told me "you will do more harm than any ten naturalists will do good"— "I can see that you have already corrupted and half-spoiled Hooker"(!!)' (p. 177). To the 'corrupted' Hooker, he wrote in 1858 of his deep satisfaction with his thorough-going reformulation of nature: 'You cannot imagine how pleased I am that the notion of Natural Selection has acted as a purgative on your bowels of immutability. Whenever naturalists can

look at species changing as certain, what a magnificent field will be
open,—on all the laws of variation,—on the genealogy of all living
beings,—on their line of migration &c &c.' (p. 193).

Darwin expresses his commitment to naturalistic, materialistic explanation over a full range of subjects — as in this perceptive method-
ological comment (1850) to his cousin W. D. Fox about non-standard
medicine: 'no one knows in disease what is the simple result of nothing being done, as a standard with which to compare Homœopathy
& all other such things. It is a sad flaw, I cannot but think in my
beloved Dr Gully [from whom Darwin took the 'water cure' with
limited results], that he believes in everything— when his daughter
was very ill, he had a clair-voyant girl to report on internal changes, a
mesmerist to put her to sleep—an homœopathist, viz D^r. Chapman;
& himself as Hydropathist! & the girl recovered' (p. 116).

Darwin expresses his social and political liberalism most forcefully
in his views on race. He does not doubt the superiority of British
culture and the rudeness of savage life, but he regards all people as
'improvable' to his preferred standard — and he loathes slavery and
any other system that would stymie such advance.

As a young man on the *Beagle*, Darwin was shocked by the 'primitive' Fuegians of southernmost South America, but he was also fas-
cinated and convinced of their brotherhood with all other humans.
He wrote to Henslow in 1833: 'The Fuegians are in a more miserable
state of barbarism, than I had expected ever to have seen a human
being.— In this inclement country, they are absolutely naked . . . I do
not think any spectacle can be more interesting, than the first sight
of Man in his primitive wildness' (p. 27). Two years later, the work of
missionaries in Tahiti confirms his views on potential 'improvement'
for all to a common high standard (and also expresses the paternal-
ism then so prevalent among racial liberals): 'The Missionaries have
done much in improving their moral character & still more in teaching them the arts of civilization . . . Europæns may here, amongst
men who, so lately were the most ferocious savages probably on the
face of the earth, walk with as much safety as in England' (p. 49).

Darwin particularly loathed slavery and the misuse of scientific
evidence in its support. In 1850, he wrote of Agassiz's advocacy of
polygeny — the doctrine that human races represent separate species:
'I wonder whether the queries . . . about the specific distinctions of
the races of man are a reflexion from Agassiz's Lectures in the U.S.
in which he has been maintaining the doctrine of several species,

—much I daresay, to the comfort of the slave-holding Southerns' (p. 115).

Darwin also supports a wide range of social reforms in the liberal mode — including educational reform with less emphasis upon the classics and rote learning ('What a grand step it would be to break down the system of eternal classics, & nothing but classics' (p. 103)); and legislation against child labor. He praises his sister Susan's fight against the common practice of using young boys as chimney sweeps: 'the brutal Shropshire Squires are as hard as stone to move . . . It makes one shudder to fancy of one's own children at 7 years old being forced up a chimney—to say nothing of the consequent loathsome disease, & ulcerated limbs, & utter moral degradation' (p. 125).

The third factor, Darwin's conservative lifestyle and social preferences, may best be glimpsed in his role of paterfamilias, giving advice to his children. He lectures his son William on the transcendent importance of good manners: 'You will surely find that the greatest pleasure in life is in being beloved; & this depends almost more on pleasant manners, than on being kind with grave & gruff manners. You are almost always kind and only want the more easily acquired external appearance. Depend upon it, that the only way to acquire pleasant manners is to try to please *everybody* you come near, your schoolfellows, servants & everyone' (p. 122).

And although Darwin supported educational reform in principle, he could not quite bring himself to 'experiment' with his own children. He considers alternatives and even visits several schools: 'We are at present very full of the subject of schools; I cannot endure to think of sending my Boys to waste 7 or 8 years in making miserable Latin verses' (p. 115). But he demurs in the same letter to his cousin W. D. Fox: 'I feel that it is an awful experiment to depart from the usual course, however bad that course may be.' Darwin eventually capitulates to tradition: 'No one can more truly despise the old stereotyped stupid classical education than I do, but yet I have not had courage to break through the trammels. After many doubts we have just sent our eldest Boy to Rugby' (p. 124).

Partly as good strategy, but partly led by these conservative tendencies as well, Darwin chooses to avoid all talk of human evolution in the *Origin of species*. He writes to A. R. Wallace: 'You ask whether I shall discuss "man";— I think I shall avoid whole subject, as so surrounded with prejudices, though I fully admit that it is the highest & most interesting problem for the naturalist' (p. 185).

All these features of intellect and feeling join with a fundamental trait of character — Darwin's toughness, zeal, perseverance, and constant enthusiasm — to establish the context of his preeminent achievements. Mere intellect is never enough to forge a scientific revolution. Darwin also possessed the requisite traits of character that make a person both persist and believe in himself — and he also had the good luck of personal wealth, powerful friends, and high status in a stratified society that offered particular advantages to the fortunate.

The post-*Beagle* letters of this book provide a wonderful chronology of Darwin's maximally varied accomplishments — but all focussed on his evolutionary reconstruction of life. Three chronological periods of intense activity precede his composition of the *Origin of species*. From the later 1830s to the early 1840s, Darwin prepares his *Beagle* specimens for publication, writes the several, mainly geological volumes inspired by the voyage, and, most crucially to later judgment, floats some hints of his evolutionary ideas after developing the theory of natural selection in 1838. Some of these hints are opaque to say the least, but the famous 5 July 1844 letter of explicit and elaborate instruction to his wife Emma shows how clearly Darwin understood the magnitude of what he had accomplished — and how much he cared for recognition of his originality: 'I have just finished my sketch of my species theory. If, as I believe that my theory is true & if it be accepted even by one competent judge, it will be a considerable step in science. I therefore write this, in case of my sudden death, as my most solemn & last request, which I am sure you will consider the same as if I legally entered in my will, that you will devote 400£ to its publication & further will yourself, or through Hensleigh, take trouble in promoting it . . . With respect to Editors.— M^r. Lyell would be the best if he would undertake it: I believe he w^d find the work pleasant & he w^d learn some facts new to him' (p. 82).

In a second phase, from the mid 1840s to the early 1850s, Darwin labors interminable hours to produce his four great monographs on the taxonomy of living and fossil barnacles, his most thorough work in biological research. The many letters on this project are particularly revealing in showing Darwin's tenacity, and both the joys and intense frustrations that such long and complexly detailed work must inspire.

At times he almost despairs of ever finishing, as in this comment to Hooker on 3 February 1850: 'I have now for a long time been at work at the fossil cirripedes, which take up more time even than the recent;—confound & exterminate the whole tribe; I can see no end to my work' (p. 114). In an earlier letter to Hooker, on 12 October 1849, Darwin makes a similar complaint about taxonomic legalisms, while stating his pleasure in empirical work on anatomy. He is now in better humor, and even chides Hooker for inconstancy of preference: 'As long as I am on anatomy I never feel myself in that disgusting, horrid cui bono inquiring humour. What miserable work, again, it is searching for priority of names . . . My chief comfort is, that the work must be sometime done, & I may as well do it, as anyone else . . . By the way, you say in your letter that you care more for my species work than for the Barnacles; now this is too bad of you, for I declare your decided approval of my plain Barnacle work over theoretic species work, had very great influence in deciding me to go on with former & defer my species-paper' (p. 111).

I was most intrigued by a fascinating comment to Hooker on 10 May 1848, showing the intimate relationship between Darwin's barnacle work and his (still hidden) evolutionary world view — and so well illustrating Darwin's favourite principle that all good empirical observation must be informed by theoretical views. Note also Darwin's humor and confidence at the end: 'I have lately got a bisexual cirripede, the male being microscopically small & parasitic within the sack of the female; I tell you this to boast of my species theory, for the nearest & closely allied genus to it is, as usual, hermaphrodite, but I had observed some minute parasites adhering to it, & these parasites, I now can show, are supplemental males, the male organs in the hermaphrodite being unusually small . . . I never sh$^\text{d}$. have made this out, had not my species theory convinced me, that an hermaphrodite species must pass into a bisexual species by insensibly small stages, & here we have it, for the male organs in the hermaphrodite are beginning to fail, & independent males ready formed. But I can hardly explain what I mean, & you will perhaps wish my Barnacles & Species theory al Diabolo [to the devil] together. But I don't care what you say, my species theory is all gospel' (p. 101).

In a third phase through the 1850s, Darwin performs experiments in preparation for his 'big species book' — primarily on variation and breeding in pigeons and other animals, and on modes of natural transport that might explain the geographic distribution of organisms after evolutionary origin of each form in a single region. For this latter subject, Darwin pursued a wide range of ingenious, often quite simple, and sometimes almost obsessively detailed experiments on such topics as how long seeds might float in salt water and still germinate, whether seeds and small eggs might be transported in mud caked on birds' feet, what seeds might pass through a bird's digestive tract and survive, etc. For example, he writes to Hooker in 1856: 'Lastly I have had a partrid[g]e with 22 grains of dry earth on *one* foot, & to my surprise a pebble as big as a tare seed; & I now understand how this is possible for the bird scartches [sic] itself, & little plumose feathers make a sort of very tenacious plaister. Think of the millions of migratory quails, & it w^d. be strange if some plants have not been transported across good arms of the sea' (p. 163).

In pursuing this work and others, Darwin shamelessly made requests upon friends, colleagues, and acquaintances in all modes of life. He would ask for anything he needed, no matter how peculiar, and no matter how extensive the effort required. He always asked politely, and with profound apology in advance for requesting so much effort — but ask he did. For me, nothing else so strongly illustrates Darwin's thoroughness and utter commitment — and also (for issues of social class so permeate Darwin's biography) his sense that people of lower status could only be flattered by the attention. He asks his former *Beagle* servant Syms Covington, now living in Australia, for antipodal barnacles: 'I do not know whether you live near the sea, but if so I should be very glad if you would collect me any that adhere . . . to the coast rocks . . . You will remember that barnacles are conical little shells, with a sort of four-valved lid on the top' (p. 107).

Darwin sends endless questions to W. B. Tegetmeier, the leading amateur expert on pigeons. He cajoles his cousin W. D. Fox about a wide range of subjects, adding as a rationale: 'I work all my friends . . . Are castrated Deer larger than ordinary Bucks? Do you know?' (p. 164). Of his friend T. C. Eyton, he begs: 'I have been trying for a year with no success to get some dace &c. Have you any & could you catch some in net. & order your kitchen maid to clean them, & you c^d. send me the whole stomach & I would sow the contents on burnt earth with every proper precaution. If ever your goodnature sh^d. lead you to send me any such rubbish; it might be put in bladder or tin foil & sent by Post' (p. 159).

Professional colleagues were by no means exempt from such requests. Darwin asks J. D. Dana for American barnacles, Asa Gray for voluminous help on botanical names and geographic ranges. He asks another colleague to translate names of domesticated varieties from Chinese encyclopedias in the British Museum. His obsessiveness with detail even leads to this request to T. H. Huxley (Darwin was right, and his correspondent wrong, in this case): 'a friend writes to me that it ought to be Geoffroy DE St. Hilaire: my memory says *no*. Will you turn to a title-page & tell me soon & forgive me asking this trouble' (p. 206).

All this leads to the epochal day in November 1859 — the publication of the *Origin of species*. It would be wrong to see everything that came before as merely a preparation (the barnacle work, for example, must be viewed primarily as *sui generis*) — and the dramas of real life meander, more than they culminate. But it would be just as wrong to deny the transcendence of the closing episode in this volume.

The basic events are well known to all who love science — Wallace's independent discovery, the 'delicate arrangement' contrived by Hooker and Lyell to publish both Wallace's paper and evidence of Darwin's earlier (though unpublished) formulation together in the Linnean Society's journal for 1858, and Darwin's subsequent decision to rush a single volume (the *Origin of species* as we know it) into print, rather than to publish the extensive treatise that he had planned. But the drama, the complexity, the moral struggles, are so much better told in Darwin's own words than through his biographers.

He writes a prophetic letter to Lyell on 3 May 1856, worrying that his long delay might lead to usurpation of his priority: 'I rather hate the idea of writing for priority, yet I certainly sh^d. be vexed if any one were to publish my doctrines before me' (p. 152). Yet he does not wish to hurry and writes to Lyell a year later: 'My everlasting species Book quite overwhelms me with work— It is beyond my powers, but I hope to live to finish it' (p. 171). He almost smells Wallace's closeness, and fires a warning shot by writing to him on 1 May 1857: 'This summer will make the 20th year (!) since I opened my first note-book, on the question how & in what way do species & varieties differ from each other.— I am now preparing my work for publication, but I find the subject so very large, that though I have written many chapters, I do not suppose I shall go to press for two years' (p. 172).

Darwin then receives Wallace's fateful paper and writes the famous letter to Lyell on 18 June 1858: 'I never saw a more striking coincidence. if Wallace had my M.S. sketch written out in 1842 he could not have made a better short abstract! Even his terms now stand as Heads of my Chapters . . . So all my originality, whatever it may amount to, will be smashed. Though my Book, if it will ever have any value, will not be deteriorated; as all the labour consists in the application of the theory' (p. 188).

By a week later, Darwin's numbed shock had given way to strategic thought, and he hinted (not at all subtly) to Lyell how his legitimate priority might be honorably preserved: 'I would far rather burn my whole book than that he or any man sh^{d} think that I had behaved in a paltry spirit. Do you not think his having sent me this sketch ties my hands? . . . If I could honourably publish I would state that I was induced now to publish a sketch . . . from Wallace having sent me an outline of my general conclusions . . . I could send Wallace a copy of my letter to Asa Gray to show him that I had not stolen his doctrine . . . My good dear friend forgive me.— This is a trumpery letter influenced by trumpery feelings' (p. 189).

So Lyell and Hooker, following Darwin's own suggestions, devised the 'delicate arrangement' while Wallace remained *hors de combat* in the East Indies. Wallace, happily, proclaimed himself satisfied with the resolution (which, in difficult circumstances, must, I think, be deemed eminently fair) — and Darwin wrote a slightly disingenuous letter (however politic) to Wallace, claiming no role whatever in the published outcome: 'I was extremely much pleased at receiving three days ago your letter to me & that to Dr Hooker. Permit me to say how heartily I admire the spirit in which they are written. Though I had absolutely nothing whatever to do in leading Lyell & Hooker to what they thought a fair course of action, yet I naturally could not but feel anxious to hear what your impression would be' (p. 197).

And the rest, as they say, is history — and current events as well, for evolution remains as powerful in its synthesis, and as controversial in its implications, as ever. We should give the last word to Charles Darwin, writing to Lyell as his book went to press: 'I cannot too strongly express my conviction of the general truth of my doctrines, & God knows I have never shirked a difficulty' (p. 205).

Introduction

The letters in this volume span the years from 1825, when Darwin was a student at the University of Edinburgh, to the end of 1859, when the *Origin of species* was published. The early letters portray Darwin as a lively sixteen-year-old medical student. Two years later he abandoned any idea of following his father in becoming a physician and transferred to Cambridge University to prepare for the ministry. His interests as an undergraduate at Cambridge, as at Edinburgh, were clearly outside the established academic curriculum. He became an enthusiastic collector of insects, and a devoted follower of the professor of botany, John Stevens Henslow, who encouraged his interest in natural history, for which no degree was then offered. Soon after Darwin took his BA degree, Henslow recommended him for the post of unoffical naturalist and companion to Robert FitzRoy, captain of HMS *Beagle*, which was being prepared for a survey voyage to South America and the Pacific.

The letters that Darwin sent to his family and to Henslow during the *Beagle*'s five-year circumnavigation of the globe contain extensive accounts of his experiences and observations. When excerpts from the letters to Henslow were made known to the learned societies in Cambridge and London, they aroused such intense interest that, by the time the *Beagle* arrived back in England in 1836, Darwin was already a well-known naturalist and an accepted member of the scientific community.

The years following his return were devoted to writing up the results of the voyage. His first book, *Journal of researches*, was to become one of the most famous travel books ever published.[1] It was followed by three volumes on the geology of the voyage, in one of which he proposed a new explanation of the formation of coral reefs that won the support of Charles Lyell, the leading English geologist of the time. With a grant of £1000 from the Treasury he superintended and edited the *Zoology of the voyage of the Beagle*, a series of monographs in nineteen parts by expert taxonomists, describing the fossil and living mammals, fish, birds, and reptiles collected during the voyage. Darwin supplied geological and geographical introductions, with notes on the habits and ranges of the species. By 1846, he had also published over twenty-five scientific papers, almost all of them directly related to the observations made during the voyage.

The prodigious quantity of work produced during these years was achieved despite several periods of an illness that was to plague Darwin for most of his life. None of his many physicians ever found the cause and no treatment provided more than temporary alleviation. To this day it remains a subject of great interest to Darwin scholars and medical historians.[2]

On 1 October 1846, Darwin, noting in his diary that he had finished the third volume of the geology of the voyage, wrote: 'Now it is 10 years since my return to England. How much time lost by illness!' On that same day, he began a description of an interesting barnacle that he had found off the coast of Chile. To understand its structure, he undertook to compare it with that of other species, and, finding that the literature on the classification of Cirripedia was chaotic and full of errors, he embarked on a study of both the fossil and living species. It was eight years before he completed what was the first taxonomic study of the entire order.

By this time, 1854, Darwin had become a family man. In January 1839, he had married his cousin Emma Wedgwood. In December of that year, William Erasmus Darwin was born, the first of ten children, two of whom died in infancy. A third, his beloved daughter Annie, died at the age of ten in 1851. The letters are an intimate chronicle of Darwin and of an affectionate family. A fascinating aspect of their close relationship is the extent to which the children, as they grew up, became active participants in Darwin's scientific work. Even at an early age he encouraged them to make botanical and entomological observations for him. Though not mentioned specifically in the letters, field notes exist that record the observations made between 1854 and 1861 by five of his children, on the flight routes of male humble-bees.

As noted above, almost all of Darwin's published work up to this time was the direct result of the *Beagle* voyage. Another, and eventually more important result, had not yet led to any published account, nor did it receive much direct mention in his letters. This was his work on what he called 'the species problem'.

On the last leg of the homeward journey, as Darwin organised his notes on the Galápagos birds, it struck him that the indigenous mocking-birds were closely allied to species on the mainland, and yet were distinct and unique to that archipelago. In London, the similarity of the fossils of extinct mammals he had found in South America to some of the living animals, and John Gould's naming of thirteen species of finches collected in the Galápagos impelled him to start investigating the possibility of evolution. His diary records his memory of the momentous decision in 1837:

> In July opened first note Book on 'transmutation of Species'—
> Had been greatly struck from about month of previous March
> —on character of S. American fossils—& species on Galapagos
> Archipelago.— These facts origin (especially latter) of all my
> views.[3]

The notebooks that he filled in the years that followed record an extraordinary amount of reading and collecting of facts on variations in plants and animals, with speculations on how species might have arisen. In September 1838, reading Thomas Malthus's *Essay on the principle of population* (London, 1826), he found a clue: in the competition for food, any variation that gave the slightest advantage would count in the struggle for survival. From then on his researches were guided by this hypothesis, which he named 'natural selection'.

The letters show that Darwin was not as secretive about his species doubts as has been commonly thought. Between 1838 and 1857, he told at least ten of his correspondents that he was investigating the mutability of species. As early as 14 September 1838, before reading Malthus, he wrote to Lyell about 'the delightful number of new views that have been coming in, thickly & steadily, on the classification & affinities & instincts of animals—bearing on the question of species— note book, after note book has been filled, with facts, which begin to group themselves clearly under sub laws.'[4]

In the years that followed, numerous letters were concerned with this search for data relevant to the species question, though without any direct mention of the theory behind the search. Correspondents all over the globe were plied with questions and requests for facts and specimens. His most important source of both information and critical discussion of his theory was, however, close at hand in the person of Joseph Dalton Hooker. Hooker had recently returned from an expedition to Antarctica to become assistant to his father, the director of the Royal Botanic Gardens at Kew. In his letter of 11 January 1844, Darwin revealed to Hooker that he thought he had found the mechanism of transmutation. There followed a lengthy and voluminous exchange of letters in which a close friendship developed, and Hooker became deeply involved in Darwin's work as counsellor, critic, and, increasingly, as a collaborator in the construction of the case for mutability and natural selection.

In September 1855, an article by Alfred Russel Wallace appeared with the title, 'On the law which has regulated the introduction of new species'.[5] Wallace had been studying the geographical distribution of animals and plants in Malaysia and had concluded that every species had 'come into existence coincident both in time and space with a pre-existing closely allied species'. To Charles Lyell this was a warning that Wallace might be on the track of something close to Darwin's theory, and he urged his friend to publish, lest he be forestalled. Darwin, somewhat reluctantly, began to write 'a sketch', but found that, in order to present a convincing proof of his theory, a much longer work was required. He had written a quarter of a million words in ten and a half chapters of what he came to call his 'big book', when, in June 1858, he received the famous letter from Wallace in which was enclosed a manuscript describing his own independent discovery of natural selection. Lyell and Hooker, to salvage the twenty years of Darwin's work, proposed that Wallace's manuscript be published jointly with some excerpts of Darwin's earlier, though unpublished, writing on the subject. The joint paper was read at the Linnean Society on 1 July 1858.[6] A few weeks later, Darwin set to work on a condensation of his long manuscript. It too grew beyond the limits of any learned journal, despite the omission of much supporting data and sources. The result was the *Origin of species*.

Editor's Note

The letters in this volume have been selected from the first seven volumes of the *The Correspondence of Charles Darwin* published by Cambridge University Press (1985–91). I have sought to make the selection representative of the larger work and to provide a trustworthy portrayal of Darwin's mind, personality, and method of work as well as an account of the important stages of his development from a student to the author of the work that has transformed our understanding of nature and mankind.

The texts of the selected letters retain the spelling and punctuation of the originals. Editorial interpolations are placed in square brackets, and all editorial omissions are indicated by ellipses. Darwin himself occasionally used ellipses, and these have been indicated in the endnotes.

Letters never intended for publication are likely to contain references that require explanation. The notes here appended are selective, taken or adapted from the fully annotated *Correspondence*, with the interests of the general reader in mind. The notes are followed by a biographical list of the individuals mentioned in the letters, as well as a bibliographical note on Darwin's writings from the years covered by this volume.

In the headings and notes throughout the volume, Darwin is referred to as CD.

Finally, I wish to express my thanks to Sarah Benton, Heidi Bradshaw, and Sarah Lavelle of the Darwin Correspondence editorial staff, and to Josie Dixon of Cambridge University Press for their invaluable assistance in bringing out this volume.

Typographical Note

In this book the editor's commentaries are set in type smaller than the main letter text; extracts from Darwin's writings are set in italics.

The Cambridge University Library computer was used for initial work on this edition and for proof production, and thanks are due to the Automation staff of Cambridge University Library for their support and assistance.

The text of this book is set in Monotype Baskerville™. Camera-ready copy was produced at the Oxford University Computing Service on a Monotype Prism PS Plus PostScript imagesetter, with the assistance of Stephen Miller and the Oxford operations staff.

Prologue

The first letter in this volume was written when Darwin was sixteen. The following provide some details of his life up to that point.

1809
Charles Robert Darwin was born on 12 February, second son and fifth of six children of Robert Waring Darwin, physician of Shrewsbury, and Susannah, daughter of Josiah Wedgwood I, master-potter of Staffordshire.

1817
Attended the Reverend George Case's school in the spring. On 15 July, Darwin's mother died, aged 52.

1818
Entered Shrewsbury School as a boarding student.

1822
Began to assist his older brother Erasmus with chemistry.

1823
Collected minerals, plants, insects.

1825
Left Shrewsbury School, 17 June. In October arrived in Edinburgh with Erasmus to study medicine.

Edinburgh

['*As I was doing no good at school, my father wisely took me away at a rather earlier age than usual, and sent me (October 1825) to Edinburgh University, with my brother where I stayed for two years*' (*Autobiography*, p. 46.)]

To Robert Waring Darwin   [23 October 1825]

[Edinburgh]
Sunday morning.

My dear Father

As I suppose Erasmus has given all the particulars of the journey I will say no more about it, except that alltogether it has cost me 7 pounds— We got into our lodgings yesterday evening, which are very comfortable & near the College— Our Landlady, by name M^{rs}. Mackay, is a nice clean old body, and exceedingly civil & attentive— She lives in "11 Lothian Street Edinburg" & only four flights of steps from the ground floor which is very moderate to some other lodgings that we were nearly taking— The terms are 1£—6^s. for two very nice & *light* bedrooms & a nice sitting room; by the way, light bedrooms are very scarce articles in Edinburg, since most of them are little holes in which there is neither air or light. We called on D^r. Hawley the first morning, whom I think we never should have found had it not been a good natured D^r. of Divinity who took us into his Library & showed us a map, & gave us how find him: Indeed all the Scotchmen are so civil and attentive, that it is enough to make an Englishman ashamed of himself—

I should think D^r. Butler or any other fat English divine would take two utter strangers into his library and show them the way! When at last we found the Doctor & having made all the proper speeches on both sides we all three set out and walked all about the town; which we admire excessively; indeed Bridge Street is the most extraordinary thing I ever saw, and when we first looked over the sides we could hardly believe our eyes, when, instead of a fine river we saw a stream of people—

1

We spend all our mornings in promenading about the town, which we know pretty well, and in the Evenings we go to the play to hear Miss Stephens, which is quite delightful. She is very popular here, being encored to such a degree that she can hardly get on with the play— On Monday we are going to Der F^{r}.[1] (I do not know how to spell the rest of the word)— Before we got into our lodgings we were staying at the Star Hotel in Princes St. . . .

The introductory lectures begin next Wednesday, and we were matriculated for them on Saturday: we pay 10^{s}. & write our names in a book, & the ceremony is finished; but the Library is not free to us till we get a ticket from a Professor—

We have just been to church and heard a sermon of only 20 minutes. I expected from Sir Walter Scott's account, a soul-cutting discourse of 2 hours & a half—

I remain Y^{r}. affectionate son | C. Darwin.

To Caroline Darwin 6 January 1826

Edinburgh.

Jan. 6th. | 1826—

My dear Caroline,

Many thanks for your very entertaining letter, which was a great relief after hearing a long stupid lecture from Duncan on Materia Medica— But as you know nothing either of the Lecture or Lecturers, I will give you a short account of them.— D^{r}. Duncan is so very learned that his wisdom has left no room for his sense, & he lectures, as I have already said, on the Materia Medica, which cannot be translated into any word expressive enough of its stupidity. These few last mornings, however, he has shown signs of improvement & I hope he will "go on as well as can be expected." His lectures begin at eight in the morning.— D^{r}. Hope begins at ten o'clock, & I like both him & his lectures *very* much. (After which Erasmus goes to M^{r}. Lizars on Anatomy", who is a charming Lecturer) At 12, the Hospital, after which *I* attend Munro on Anatomy— I dislike him & his Lectures so much that I cannot speak with decency about them. He is so dirty in person & actions.— Thrice a week we have what is called Clinical Lectures, which means lectures on the sick people in the Hospitals—these I like *very* much.— I said this account should be short, but I am afraid it has been too long like the Lectures themselves.—

I will be a good *boy*, and tell something about Johnson again (not but what I am very much surprised that Papa should so forget himself as to call me, a Collegian in the University of Edinburgh, a boy.) he has changed his lodgings for the third time, he has got very cheap ones, but I am afraid it will not answer, for they must make up by cheating.— I hope you like Erasmus' official news, he means to begin every letter so.— You mentioned in your letter that Emma was staying with you, if she is not gone ask her to tell Jos. that I have not succeeded in getting any [titanium], but that I will try again. Tell Katty and Susan I shall be very grateful if they will write to me, it is so pleasant receiving letters; and I hope, although our correspondence has begun late, you will send me many more nice affecting letters about dear little black nose. Erasmus thinks I shall have more pleasure in seeing it than all the rest of the families put together. You seem to hold the same opinion with regard to my dear little nephew.— I want to know how old I shall be next Birthday. I believe 17, & if so I shall be forced to go abroad for one year since it is necessary that I shall have completed my 21st. year before I take my degree. Now you have no business to be frowning & puzzling over this letter for I did not promise to write a good hand to you.

I remain your af— dear Caroline, | C. Darwin.

Love to Papa & tell him I am going to write to him in a few days—

To Susan Darwin 29 January [1826]

Edinburgh.

Jan 29.

My dear Susan

The whole family have been so very good in writing to me so often that I do not know whom to begin to thank first, so to save trouble I return my humble thanks to you all, from my Father down to little Kitty.— The Gaieties of Edinburgh are now just beginning, last week there was an Assembly, & shortly there will be another. Erasmus & ⟨ ⟩ intended to have gone to the first, but mean to make it up by going to the next.— We also have been very dissipated.— We dined at D^r. Hawley's on Saturday, & had a very pleasant party, after which we went to the Theatre, with a M^r. Greville I believe a relation of the great Botanist, D^r. Greville. D^r. Hawley has procured some information about my Fathers questions & will write it shortly to him. Next Friday we are going to the old D^r. Duncan, & I hope it will be a pleasanter party than the last; which a very specimen of stupidity. What an extraordinary old man he is, now being past 80,

& continuing to lecture. D.ʳ Hawley hints that he is rapidly failing. I
have been most shockingly idle, actually reading two novels at once.
a good scolding would do me a vast deal of good, & I hope you will
send one of your most severe one's.— What an entertaining book
Granby is;[2] do you remember Lady Harriet talking about inhaling
⟨Ni⟩tric Oxide? Johnson has actually done it, & describes the effects
as the most intense pleasure he ever felt. We both mean to get tipsey
in the Vacation.—. The old M.ʳ Wedgwood, I see in Ure's Chem.
Dic., did nothing else but hold his nose & kick. It occasionally brings
on fainting. Erasmus knows a man in Cambridge, who when in that
state had the faculty of hearing, but not of motion or speech & to his
horror, heard them consulting whether they should open the Temporal Artery. . . . I am going to learn to stuff birds, from a blackamoor
I believe an old servant of D.ʳ Duncan: it has the recommendation
of cheapness, if it has nothing else, as he only charges one guinea,
for an hour every day for two months.

I rem.n— | Charles Darwin . . .

To Caroline Darwin 8 April [1826]

[Edinburgh]
April 8th

My dear Caroline

   I dare say I shall not be able to finish this letter, but I cannot
help writing to thank you for your very nice and kind letter. It makes
me feel how very ungrateful I have been to you for all the kindness
and trouble you took for me when I was a child. Indeed I often
cannot help wondering at my own blind Ungratefulness. I have tried
to follow your advice about the Bible, what part of the Bible do you
like best? I like the Gospels. Do you know which of them is generally
reckoned the best? Do write to me again soon, for you do not know
how I like receiving such letters as yours. The weather has been very
pleasant for these few last days, but nevertheless I long to return
very much. D.ʳ Hope has been giving some very good Lectures on
Electricity &c. and I am very glad I stayed for them. The Classes are
beginning to thin. I think I shall stay about nine days or a fortnight
longer. But as my money will not last out for the latter period I should
be much obliged if my Father would send up a £5 or £10 pound
bill. I should also be much obliged if it could be sent up **directly**:
as you will not receive this letter for three days, and I get the answer
in another three.

I hope Eras. has got his books safely, let me know in the next letter, as I mean to go as far as Glasgow by the canals and from thence on "terra firma" to Shrewsbury, but sending my books per sea. How shockingly untidy this letter is, but I think you have a sympathy for untidiness, not that I mean to say that yours are half as bad as this. I will send my next and last John Bull[3] to Shrewsbury for your edification. I hope you received the last and studied it well. I said in the beginning of this very tidy epistle that I never should be able to finish it, which is now verified.

Love to all. Do not show this. | Your affect., Dear Caroline | Charles Darwin . . .

[No letters survive from CD's second year at Edinburgh. CD's brother, Erasmus Alvey Darwin, had decided to pursue his medical studies in London; CD, left on his own, found medicine less and less to his liking, and engaged in broader studies in natural history. A notebook of zoological observations made in March 1827 records his observations on two marine invertebrates, *Flustra* and *Pontobdella muricata*, which he reported to the undergraduate Plinian Natural History Society of the University on 27 March 1827.]

Cambridge

['*After having spent two sessions in Edinburgh, my father perceived or he heard from my sisters, that I did not like the thought of being a physician, so he proposed that I should become a clergyman. He was very properly vehement against my turning an idle sporting man, which then seemed my probable destination.*' (*Autobiograpy*, p. 56.)

This decision made, CD entered Christ's College, Cambridge early in 1828 to earn the necessary degree. There he met his second cousin, William Darwin Fox, who introduced him to entomology.]

To William Darwin Fox 12 [June 1828]

Friday 12th—

My dear Fox

I am dying by inches, from not having any body to talk to about insects:—my only reason for writing, is to remove a heavy weight from my mind, so now you must understand, what you will perceive before you come to the end of this; that I am writing merely for my own pleasure & not your's.— I have been very idle since I left Cambridge in every possible way & amongst the rest in Entomology. I have however captured a few insects, about which I am much interested: My sister has made rough drawings of three of them: I. fig: is I am nearly sure, the same insect as H⟨oa⟩r, of Queens took in a Willow tree, & which Garland did not know. I took [it] under the bark of a rail, was very active, striking looking insect, took 3 specimens I think this is an admirable prize

II. fig: is an *extremely* common insect; of the family of scarabidæ. Do you know it's name?—

III. fig: A *most* beautiful Leptura(?) very like the Quadrifasciata, only the body is of the same size thoughout.— I tell you all these particulars, as I am anxious to know something about these little g⟨ ⟩s.— I was not *fully* aware of your extreme value before I left Cambridge. I am constantly saying "I do wish Fox was here".— And I again say, I hope you will come & pay me visit before the

6

summer is over— My Father desired me to say, that he should be at anytime most happy to see you.—

I have taken 3 species of Coccinellæ, one, the same as Hoar took in the Fens, which you said was rare, & another with 7 *white!* marks on each elytron.— I will mention, as I believe you are interested about it, that I have seen the Cocc: bipunctata (or dispar) 4 or 5 in actu coitus with a black one with *4* red marks (I believe most of the black ones you have got have 〈   〉 marks, & hence I suppose a different species) also, which is very singular, I have frequently seen two of the bipunctata's in actu.— I Have taken Clivina Collaris, fig 〈3〉 Plate III of Stephens;[1] also a beautiful copper-coloured Elater

(with Antennæ  pectin〈ate〉 like this.

Do you want any of the Byrrhus Pillula? I can get any number.— ...

Believe me my dear Fox | Yours most sincerely | Charles Darwin Shrewsbury

of a fine bluish black colour, but is not so broad as made in this drawing

I

5 tarsi

4 Tarsi

rather lighter coloured & more metallic the legs are left out.—

II

this is a very good representation

III

I. fig is more like a Pyrochroa or a very narrow Blaps than any thing I can compare it to.—

I.I. fig: be sure to give me the name of this insect

To Erasmus Alvey Darwin   21 December 1828

[Shrewsbury]

My dear Erasmus

Susan has given me this third sheet to write to you upon; but why I should do so I know not. Apologies are come much too late for such shamefully ungrateful conduct as mine has been and excuses I have none. You are become like a stranger to me. . . . After you left Cambridge I got into very nice rooms in College, far more comfortable than lodgings, as you will find when you come next to Cambridge. I imbibed your tastes about prints, and put it into practice, and have bought some very good prints, which I long for you to see. I came down from Cambridge yesterday which I left very empty. I think I shall go and see old Edinburg this X'tmas, before all my friends entirely leave it. I left Fox up in Cam. in great awe and tribulation about his degree, which is to be made very much more strict, so that they give out that at least 50 will be plucked: I live almost entirely with Fox and Entomology goes on most surprisingly. Price is in Cambridge and has got some pupils, he intends writing to you very soon. Kennedy is a tutor at St. Johns: Whitley will be either first or second next year: I see a good deal of him and like him very much. This will be but a poor finishing to Susan's letter, and must be awfully stupid. When is there any prospect of your coming back again? I long very much to see the Bachelor of Medicine again. Fanny Owen is as charming as ever and deserves your string of German epithets more than ever.[2]

My dear old Erasmus | Yours most affectionately | C. Darwin

To W. D. Fox [25 March 1830]

[Cambridge]
Thursday

My dear Fox

I am through my little Go,!!!![3] I am too much exalted to humble myself by apologising for not having written before.— But, I assure you before I went in & when my nerves were in a shattered & weak condition, your injured person often rose before my eyes & taunted me with my idleness. But I am through through through. I could write the whole sheet full, with this delightful word.— I went in yesterday, & have just heard the joyful news.— I shall not know for a week, which class I am in.— The whole examination is carried on in a different system. It has one grand advantage, being over in one day. They are rather strict; & ask a wonderful number of questions:

And now I want to know something about your plans: *of course* you intend coming up here: what fun we will have together, what beetles we will catch, it will do my heart good to go once more together to some of our old haunts: I have two very promising pupils in Entomology, & we will make regular campaigns into the Fens; Heaven protect the beetles & Mr Jenyns, for we wont leave him a pair in the whole country. My new cabinet is come down & a gay little affair it is.— And now for the time, I think I shall go for a few days to Town, to hear an Opera & see M^r. Hope; not to mention my brother also whom I should have no objection to see.— If I go pretty soon, you can come afterwards but if you will settle your plans definitely, I will arrange mine. So send me a letter by return of post:— And I charge you let it be favourable, that is to say come directly. . . .

I long to see you again, & till then | My dear good old Fox | I am yours most sincerely | C. Darwin

[CD took his examination for the BA degree on 22 January 1831, ranking tenth in the 168 undergraduates who took pass degrees. In his final two terms of residence, his mentor and friend, Professor John Stevens Henslow, encouraged him to take up geology, a subject he had given up as hopelessly dull in his Edinburgh undergraduate years. In July he was preparing for a geological expedition to Wales with Adam Sedgwick, Woodwardian Professor of Geology at Cambridge.]

To John Stevens Henslow [11 July 1831]

Shrewsbury
Monday

My dear Sir

I should have written to you sometime ago, only I was determined to wait for the Clinometer: & I am very glad to say I think it will answer admirably: I put all the tables in my bedroom, at every conceivable angle & direction I will venture to say I have measured them as accurately as any Geologist going could do.— It cost 25^s. made of wood, but the lid with plate of brass graduated. . . . I have been working at so many things: that I have not got on much with Geology: I suspect, the first expedition I take, clinometer & hammer in hand, will send me back very little wiser & good deal more puzzled than when I started.— As yet I have only indulged in hypotheses; but they are such powerful ones, that I suppose, if they were put into action but for one day, the world would come to an end.— I have not heard from Prof: Sedgwick, so I am afraid he will not pay the seven formations a visit.—[4] I hope & trust you did your best to urge him:— . . . I hope you continue to fan your Canary ardor: I read & reread Humboldt,[5] do you do the same, & I am sure nothing will prevent us seeing the Great Dragon tree.— Would you tell L. Jenyns,

that his magnificent present of Diptera has not been wasted on me
Would you ask him how he manages Diptera when too small for a
pin to go through.— ...

Excuse all the trouble I am giving you, & Believe me my dear Sir
| Yours ever most sincerely | Chas. Darwin

Eyton begs to be most kindly remembered to you.— his mind is
in a fine tropical glow.—

The Offer

[On 24 August, Henslow wrote to CD: 'I have been asked by [George] Peacock ... to recommend him a naturalist as companion to Capt Fitzroy employed by Government to survey the S. extremity of America.— I have stated that I considered you to be the best qualified person I know of who is likely to undertake such a situation— I state this not on the supposition of y^r being a *finished* Naturalist, but as amply qualified for collecting, observing, & noting any thing worthy to be noted in natural History.']

To J. S. Henslow   30 [August 1831]

Shrewsbury
Tuesday 30th.—

My dear Sir

M^r. Peacocks letter arrived on Saturday, & I received it late yesterday evening.— As far as my own mind is concerned, I should I think, *certainly* most gladly have accepted the opportunity, which you so kindly have offered me.— But my Father, although he does not decidedly refuse me, gives such strong advice against going.—that I should not be comfortable, if I did not follow it.— My Fathers objections are these; the unfitting me to settle down as a clergyman.— my little habit of seafaring.— the *shortness of the time* & the chance of my not suiting Captain Fitzroy.— It is certainly a very serious objection, the very short time for all my preparations, as not only body but mind wants making up for such an undertaking.— But if it had not been for my Father, I would have taken all risks.—

What was the reason, that a Naturalist was not long ago fixed upon?— I am very much obliged for the trouble you have had about it—there certainly could not have been a better opportunity.— I shall come up in October to Cambridge, when I long to have some talk with you. . . . My trip with Sedgwick answered most perfectly.— I did not hear of poor M^r. Ramsays loss till a few days before your letter. I have been lucky hitherto, in never losing any person for whom I had any esteem or affection. My Acquaintance, although very short, was sufficient to give me those feelings in a great degree.— I can hardly

11

make myself believe he is no more.— He was the finest character I ever knew.—

Yours most sincerely | my dear Sir. Chas. Darwin ...

Even if I was to go my Father disliking would take away all energy, & I should want a good stock of that. ...

['*On the last day of August I went to Maer* [the Wedgwood family home] *where everything soon bore a different appearance. I found every member of the family so strongly on my side, that I determined to make another effort.*' (*Beagle diary*, p. 3.)]

To R. W. Darwin 31 August [1831]

[Maer]
August 31

My dear Father

I am afraid I am going to make you again very uncomfortable.— But upon consideration, I think you will excuse me once again stating my opinions on the offer of the Voyage.— My excuse & reason is, is the different way all the Wedgwoods view the subject from what you & my sisters do.—

I have given Uncle Jos, what I fervently trust is an accurate & full list of your objections, & he is kind enough to give his opinion on all.— The list & his answers will be enclosed.— But may I beg of you one favor. it will be doing me the greatest kindness, if you will send me a decided answer, yes or no.— If the latter, I should be most ungrateful if I did not implicitly yield to your better judgement & to the kindest indulgence which you have shown me all through my life.—& you may rely upon it I will never mention the subject again.— if your answer should be yes; I will go directly to Henslow & consult deliberately with him & then come to Shrewsbury.— The danger appears to me & all the Wedgwoods not great.— The expence can not be serious, & the time I do not think anyhow would be more thrown away, than if I staid at home.— But pray do not consider, that I am so bent on going, that I would for one *single moment* hesitate, if you thought, that after a short period, you should continue uncomfortable.—

I must again state I cannot think it would unfit me hereafter for a steady life.— I do hope this letter will not give you much uneasiness.— I send it by the Car tomorrow morning if you make up your mind directly will you send me an answer on the following day, by the same means.— If this letter should not find you at home,

12

I hope you will answer as soon as you conveniently can.—
I do not know what to say about Uncle Jos.' kindness, I never can
forget how he interests himself about me
    Believe me my dear Father | Your affectionate son | Charles
Darwin. ...

(1)   Disreputable to my character as a Clergyman hereafter
(2)   A wild scheme
(3)   That they must have offered to many others before me, the
      place of Naturalist
(4)   And from its not being accepted there must be some serious
      objection to the vessel or expedition
(5)   That I should never settle down to a steady life hereafter
(6)   That my accomodations would be most uncomfortable
(7). That you should consider it as again changing my profession
(8)   That it would be a useless undertaking

[Uncle Jos's letter answered each point in CD's favour, concluding with point 8:
'The undertaking would be useless as regards his profession, but looking upon
him as a man of enlarged curiosity, it affords him such an opportunity of seeing
men and things as happens to few.' The letters were sent off and CD went out
shooting.

*'About ten o'clock Uncle Jos sent me a message to say he intended going to
Shrewsbury & offering to take me with him. When we arrived there, all things
were settled & my Father most kindly gave his consent.' (Beagle diary,* p. 3.)

While at Cambridge, consulting with Henslow about the voyage, CD met Alexander Wood, a great friend of Captain FitzRoy's. Wood wrote to the Captain about
CD. The reply was disconcerting.]

To Susan Darwin [5 September 1831]

                    17 Spring Gardens London
                                    Monday
... The last letter was written in the morning. in middle of day
Wood received a letter from C. Fitzroy, which I must say was *most*
straightforward & **gentlemanlike**, but so much against my going,
that I immediately gave up the scheme.—& Henslow did the same:
saying that he thought Peacock has acted *very wrong* in misrepresenting
things so much.— I scarcely thought of going to Town, but here I
am & now for more details & much more promising ones.— Cap
Fitzroy is town & I have seen him; it is no use attempting to praise
him as much as I feel inclined to do, for you would not believe me.—
One thing I am certain of nothing could be more open & kind than

he was to me.— It seems he had promised to take a friend with him, who is in office & cannot go.—& he only received the letter 5 minutes before I came in: & this makes things much better for me, as want of room of was one of Fs greatest objections.— He offers me to go share in every thing in his cabin, if I like to come; & every sort of accomodation than I can have but they will not be numerous.— He says that nothing would be so miserable for him as having me with him if I was unformfortable, as in small vessel we must be thrown together, & thought it his duty to state every thing in the worst point of view: I think I shall go on Sunday to Plymouth to see the Vessel.— There is something most extremely attractive in his manners, & way of coming straight to the point.— If I live with him he say I must live poorly, no wine & the plainest dinners.— The scheme is not certainly so good as Peacock describes: C F. advises me not make my mind quite yet: but that seriously, he thinks it will have much more pleasure than pain for me.—

The Vessel does not sail till the 10th of October.— it contains 60 men 5 or 6 officers &c.—but is a small vessel.— it will probably be out nearly 3 years.— I shall pay to mess the same as Captain does himself 30£ per annum, & Fitzroy says if I spend including my outfitting 500 it will be beyond the extreme.— But now for still worse news, the round the world is not *certain*, but the chance, most excellent: till that point is dicided I will not be so.— And you may believe after the many changes I have made, that nothing but my reason shall dicide me.—

Fitzroy says the stormy sea is exaggerated that if I do not chuse to remain with them, I can at any time get home to England, so many vessels sail that way & that during bad weather (probably 2 months) if I like, I shall be left in some healthy, safe & nice country: that I shall alway have assistance.— that he has many books, all instrument, guns, at my service.— that the fewer & cheaper clothes I take the better.—

The manner of proceeding will just suit me. they anchor the ship & then remain for a fortnight at a place.—

I have made Cap Beaufort perfectly understand me.: he says if I start & do not go round the world: I shall have good reason to think myself deceived.— I am to call the day after tomorrow, & if possible to receive more certain instructions.— The want of room is decidedly the most serious objection: but Cap Fitz. (probably owing to Woods letter) seems determined to make me comfortable as he

14

possibly can.— I like his manner of proceeding.— He asked me at once.— "shall you bear being told that I want the cabin to myself? when I want to be alone.— if we treat each other this way, I hope we shall suit, if not probably we should wish each other at the Devil" We stop a week at the Madeira islands: & shall see most of big cities in S. America. C. Beaufort is drawing up the track through the South Sea.—

I am writing in great hurry: I do not know whether you take interest enough to excuse treble postage.— I hope I am judging reasonably, & not through prejudice about Cap. Fitz: if so I am sure we shall suit.— I dine with him to day.— I could write great deal more if I thought you liked it, & I had at present time.— There is indeed a tide in the affairs of men, & I have experienced it, & I had *entirely* given it up till 1 to day:

Love to my Father, dearest Susan | good bye, Chas. Darwin

[On 5 September 1831, Robert Fitzroy wrote his impressions of CD to Francis Beaufort:

I have seen a good deal of Mr. Darwin, to-day having had nearly two hours' conversation in the morning and having since dined with him.

I like what I see and hear of him, much, and I now request that you will apply for him to accompany me as a Naturalist. I can and will make him comfortable on board, more so perhaps than you or he would expect, and I will contrive to stow away his goods and chattels of all kinds and give him a place for a workshop.

Upon consideration, I feel confident that he will have a much wider field for his exertions than I was inclined to anticipate on Friday last; and should we even be disappointed, by giving me the means of discharging him from the Books, he might at any time return to England or follow his own inclinations in South America or elsewhere.[1]

After much delay due to outfitting and stormy weather, the *Beagle* sailed from Plymouth on 27 December 1831.]

The Voyage: South America – East Coast

To R. W. Darwin 8 February – 1 March 1832

(Brazils) | Bahia or St. Salvador

My dear Father

I am writing this on the 8th of February one days sail past St. Jago, (Cape De Verd), & intend taking the chance of meeting with a homeward bound vessel somewhere about the Equator.— The date however will tell this whenever the opportunity occurs.— I will now begin from the day of leaving England & give a short account of our progress.—

We sailed as you know on the 27th. of December & have been fortunate enough to have had from that time to the present a fair & moderate breeze: It afterward proved that we escaped a heavy gale in the Channel, another at Madeira, & another on coast of Africa.— But in escaping the gale, we felt its consequence—a heavy sea: In the Bay of Biscay there was a long & continued swell & the misery I endured from sea-sickness is far far beyond what I ever guessed at.— I believe you are curious about it. I will give all my dear-bought experience.— Nobody who has only been to sea for 24 hours has a right to say, that sea-sickness is even uncomfortable.— The real misery only begins when you are so exhausted—that a little exertion makes a feeling of faintness come on.— I found nothing but lying in my hammock did me any good.— I must especially except your receipt of raisins, which is the only food that the stomach will bear:— On the 4th of January we were not many miles from Madeira: but as there was a heavy sea running, & the Island lay to Wind ward it was not thought worth while to beat up to it.— It afterwards has turned out it was lucky we saved ourselves the trouble: I was much too sick even to get up to see the distant outline.— On the 6th in the evening we sailed into the harbour of Santa Cruz.— I now first felt even moderately well, & I was picturing to myself all the delights of fresh fruit growing in beautiful valleys, & reading Humboldts descriptions of the Islands glorious views.— When perhaps you may nearly guess

16

at our disappointment, when a small pale man informed us we must perform a strict quarantine of 12 days. There was a death like stillness in the ship; till the Captain cried "Up Jib", & we left this long wished for place.— We were becalmed for a day between Teneriffe & the grand Canary & here I first experienced any enjoyment: the view was glorious. The peak of Teneriffe. —was seen amongst the clouds like another world.— Our only drawback was the extreme wish of visiting this glorious island. . . .

From Teneriffe to S^t. Jago, the voyage was extremely pleasant.— I had a net astern the vessel, which caught great numbers of curious animals, & fully occupied my time in my cabin, & on deck the weather was so delightful, & clear, that the sky & water together made a picture.— On the 16th. we arrived at Port Praya, the capital of the Cape de Verds, & there we remained 23 days viz till yesterday the 7th. of February.— The time has flown away most delightfully, indeed nothing can be pleasanter; exceedingly busy, & that business both a duty & a great delight.— I do not believe, I have spent one half hour idly since leaving Teneriffe: St Jago has afforded me an exceedingly rich harvest in several branches of Nat: History.— I find the descriptions scarcely worth anything of many of the commoner animals that inhabit the Tropic.— I allude of course to those of the lower classes.— Geologising in a Volcanic country is most delightful, besides the interest attached to itself it leads you into most beautiful & retired spots.—

Nobody but a person fond of Nat: history, can imagine the pleasure of strolling under Cocoa nuts in a thicket of Bananas & Coffee plants, & an endless number of wild flowers.— And this Island that has given me so much instruction & delight, is reckoned the most uninteresting place, that we perhaps shall touch at during our voyage.— It certainly is generally very barren.—but the valleys are more exquisitely beautiful from the very contrast:— It is utterly useless to say anything about the Scenery.— it would be as profitable to explain to a blind man colours, as to person, who has not been out of Europe, the total dissimilarity of a Tropical view.— Whenever I enjoy anything I always either look forward to writing it down either in my log Book (which increases in bulk) or in a letter.— So you must excuse raptures & those raptures badly expressed.—

I find my collections are increasing wonderfully, & from Rio I think I shall be obliged to send a Cargo home.— All the endless delays, which we experienced at Plymouth, have been most fortunate, as I verily believe no person ever went out better provided for collecting & observing in the different branches of Natural hist.— In a multitude of counsellors I certainly found good.— I find to my great surprise that a ship is singularly comfortable for all sorts of work.— Everything is so close at hand, & being cramped, make one so methodical, that in the end I have been a gainer.—

I already have got to look at going to sea as a regular quiet place, like going back to home after staying away from it.— In short I find a ship a very comfortable house, with everything you want, & if it was not for sea-sickness the whole world would be sailors.— I do not think there is much danger of Erasmus setting the example, but in case there should be, he may rely upon it he does not know one tenth of the sufferings of sea-sickness.— I like the officers much more than I did at first.—especially Wickham & young King, & Stokes & indeed all of them.— The Captain continues steadily very kind & does everything in his power to assist me.— We see very little of each other when in harbour, our pursuits lead us in such different tracks..— I never in my life met with a man who could endure nearly so great a share of fatigue.— He works incessantly, & when apparently not employed, he is thinking.— If he does not kill himself he will during this voyage do a wonderful quantity of work.— I find I am very well & stand the little heat we have had as yet as well as any-body.— We shall soon have it in real ernest.— We are now sailing for Fernando Norunho off the coast of Brazil.—where we shall not stay very long, & then examine the shoals between there & Rio, touching perhaps at Bahia:— I will finish this letter, when an opportunity of sending it occurs.—

Feb 26th. about 280 miles from Bahia.— On the 10th we spoke the packet Lyra on her voyage to Rio. I sent a short letter by her to be sent to England on first opportunity.— We have been singularly unlucky in not meeting with any homeward bounds vessels, but I suppose Bahia we certainly shall be able to write to England.— Since writing the first part of letter nothing has occurred except crossing the Equator & being shaved.— This most disagreeable operation consists of having your face rubbed with paint & tar, which forms a lather for a saw which represents the razor & then being half drowned in a sail filled with salt water.— About 50 miles North of the line, we touched at the rocks of St Paul.— this little speck (about $\frac{1}{4}$ of a mile across) in the atlantic, has seldom been visited.— It is totally barren, but is covered by hosts of birds.— they were so unused to men that we found we could kill plenty with stones & sticks.— After remaining some hours on the island, we returned on board with the boat loaded with our prey.— From this we went to Fernando Noronha, a small island where the Brazilians send their exiles.— The landing there was attended with so much difficulty owing a heavy surf, that the Captain determined to sail the next day after arriving.— My one day on shore was exceedingly interesting. the whole island is one single wood so matted together by creepers, that it is very difficult to move out of beaten path.— I find the Nat: History of all these unfrequented spots most exceedingly interesting, especially the geology.

I have written this much in order to save time at Bahia.— Decidedly the most striking thing in the Tropics is the novelty of the vegetable forms.— Cocoa Nuts could well be imagined from drawings if you add to them a graceful lightness, which no European tree partakes of.— Bananas & Plantains, are exactly the same as those in hothouses: the acacias or tamarinds are striking from blueness of their foliage: but of the glorious orange trees no description no drawings, will give any just idea: instead of the sickly green of our oranges, the native ones exceed the portugal laurel in the darkness of their tint & infinitely exceed it in beauty of form.—

Cocoa-nuts, Papaws.—the light-green Bananas & oranges loaded with fruit generally surround the more luxuriant villages.— Whilst viewing such scenes, one feels the impossibility than any description should come near the mark,— much less be overdrawn.—

March 1st. Bahia or St. Salvador.— I arrived at this place on the 28th of Feb & am now writing this letter after having in real earnest strolled in the forests of the new world.— "No person could imagine anything so beautiful as the antient town of Bahia; it is fairly embosomed in a luxuriant wood of beautiful trees.—& situated on a steep bank overlooks the calm waters of the great bay of All Saints.— The houses are white & lofty, & from the windows being narrow and long have a very light & elegant appearance   Convents, porticos & public buildings vary the uniformity of the houses: the bay is scattered over with large ships. in short & what can be said more it is one of the finest views in the Brazils".— (copied from my journal) But the exquisite glorious pleasure of walking amongst such flowers, & such trees cannot be comprehended, but by those who have experienced it.— Although in so low a Latitude the weather is not disagreeably hot, but at present it is very damp, for it is the rainy season.— I find the climate as yet agrees admirably with me: it mak⟨es⟩ one long to live quietly for some time in suc⟨h⟩ a country.— If you really want to have a ⟨notion⟩ of tropical countries, *study* Humboldt.— Skip th⟨e⟩ scientific parts & commence after leaving Teneriffe.— My feelings amount to admiration the more I read him.— Tell Eyton (I find I am writing to my sisters!) how exceedingly I enjoy America & tha⟨t⟩ I am sure it will be a great pity if ⟨he⟩ does not make a start.— This letter will go ⟨on⟩ the 5th & I am afraid will be some time before it reaches you.— it must be a warning, how in other parts of the world, you may be a long time without hearing from.— A year might by accident thus pass.—

About the 12th we start for Rio, but remain some time on the way in sounding the Albrolhos shoals. Tell Eyton, as far as my experience goes let him study Spanish French, Drawing & Humboldt. I do sincerely hope to hear of (if not to see him), in S America.— I look forward to the letters in Rio. till each one is acknowledged mention its date in the next: We have beat all the ships in mæneuvering, so much so that commanding officer says we need not follow his example, because we do everything better than his great ship.— I begin to take great interest in naval points, more especially now, as I find they all say, we are the No 1 in South America.— I suppose the Captain is a most excellent officer.— It was quite glorious to day how we beat the Samarang in furling sails: It is quite a new thing for a "sounding ship" to beat a regular man of war.— And yet the Beagle is not at all a particular ship: Erasmus will clearly perceive it, when he hears that in the night I have actually sat down in the sacred precincts of the Quarter deck.— You must excuse these queer letters, & recollect they are generally written in the evening after my days work.— I take more pains over my Log Book.—so that eventually you will have a good account of all the places I visit.—

Hitherto the voyage has answered **admirably** to me, & yet I am now more fully aware of your wisdom in throwing cold water on the whole scheme: the chances are so numerous of it turning out quite the reverse.— to such an extent do I feel this that if my advice was asked by any person on a similar occasion I should be very cautious in encouraging him.— I have not time to write to any body else: so send to Maer to let them know that in the midst of the glorious tropical scenery I do not forget how instrumental they were in placing me there.— I will not rapturize again: but I give myself great credit in not being crazy out of pure delight.—

Give my love to every soul at home, & to the Owens

I think ones affections, like other good things, flourish & increase in these tropical regions.— The conviction that I am walking in the new world, is even yet marvellous in my own eyes, & I daresay it is little less so to you, the receiving a letter from a son of yours in such a quarter: Believe me, my dear Father Your most affectionate son | Charles Darwin ...

I find after the first page I have been writing to my sisters

To Caroline Darwin 2–6 April 1832

My dear Caroline.—

... Rio de Janeiro. April 5th.— I this morning received your letter of Decr 31 & Catherines of Feb 4th.— We lay to during last night, as the Captain was determined we should see the harbor of Rio & be ourselves seen in broard daylight.— The view is magnificent & will improve on acquaintance; it is at present rather too novel to behold Mountains as rugged as those of Wales, clothed in an evergreen vegetation, & the tops ornamented by the light form of the Palm.— The city, gaudy with its towers & Cathedrals is situated at the base of these hills, & command a vast bay, studded with men of war the flags of which bespeak every nation.—

We came, in first rate style, alongside the Admirals ship, & we, to their astonishment, took in every inch of canvass & then immediately set it again: A sounding ship doing such a perfect mæneuovre with such certainty & rapidity, is an event hitherto unknown in that class.— It is a great satisfaction to know that we are in such beautiful order & discipline.— In the midst of our Tactics the bundle of letters arrived.— "Send them below," thundered Wickham "every fool is looking at them & neglecting his duty" In about an hour I succeded in getting mine, the sun was bright & the view resplendent; our little ship was working like a fish; so I said to myself, I will only just look at the signatures:, it would not do; I sent wood & water, Palms & Cathedrals to old Nick & away I rushed below; there to feast over the thrilling enjoyment of reading about you all: at first the contrast of home, vividly brought before ones eyes, makes the present more exciting; but the feeling is soon divided & then absorbed by the wish of seeing those who make all associations dear.—

It is seldom that one individual has the power giving to another such a sum of pleasure, as you this day have granted me.— I know not whether the conviction of being loved, be more delightful or the corresponding one of loving in return.— I ought for I have experienced them both in excess.— With yours I received a letter from Charlotte, talking of parsonages in pretty countries & other celestial views.— I cannot fail to admire such a short sailor-like "splicing" match.— The style seems prevalent, Fanny seems to have done the business in a ride.— Well it may be all very delightful to those concerned, but as I like unmarried woman better than those in the blessed state, I vote it a bore: by the fates, at this pace I have no chance for the parsonage: I direct of course to you as Miss Darwin.— I own I am curious to know to whom I am writing.— Susan I suppose bears the honors of being M^{rs} J Price.— I want to write to Charlotte—& how & where to direct; I dont know: it positively is an inconvenient fashion this marrying: Maer wont be half the place it was, & as for Woodhouse, if Fanny was not perhaps at this time Mrs Biddulp, I would say poor dear Fanny till I fell to sleep.—[1] I feel much inclined to philosophize but I am at a loss what to think or say; whilst really melting with tenderness I cry my dearest Fanny why I demand, should I distinctly see the sunny flower garden at Maer; on the other hand, but I find that my thought & feelings & sentences are in such a maze, that between crying & laughing I wish you all good night.— ...

So my dearest Caroline & all of you | Good bye.— Yrs very affectionately | Chas. Darwin ...

To J. S. Henslow [23 July –] 15 August [1832]

My dear Henslow

We are now beating up the Rio Plata, & I take the opportunity of beginning a letter to you. ...

And now for an apologetical prose about my collection.— I am afraid you will say it is very small.—but I have not been idle & you must recollect that in lower tribes, what a very small show hundreds of species make.— The box contains a good many geological specimens.— I am well aware that the greater number are too small.— But I maintain that no person has a right to accuse me, till he has tried carrying rocks under a Tropical sun.— I have endeavoured to get specimens of every variety of rock, & have written notes upon all.— If you think it worth your while to examine any of them, I shall be *very* glad of some mineralogical information, especially in any numbers between 1 & 254, which include St Jago rocks.— By my Catalogue, I shall know which you may refer to.— As for my Plants, "pudet pigetque mihi". All I can say is that when objects are present which I can observe & particularize about, I cannot summon resolution to collect where I know nothing.—

It is positively distressing, to walk in the glorious forest, amidst such treasures, & feel they are all thrown away upon one.— My collection from the Abrolhos is interesting as I suspect it nearly contains the whole flowering Vegetation, & indeed from extreme sterility the same may almost be said of St Jago.— I have sent home 4 bottles with animals in spirits I have three more, but would not send them till I had a fourth.— I shall be anxious to know how they fare.— I made an enormous collection of Arachnidæ at Rio.— Also a good many small beetles in pill-boxes; but it is not the best time of year for the latter.— As I have only $\frac{3}{4}$ of a case of Diptera &c I have not sent them.— Amongst the lower animals, nothing has so much interested me as finding 2 species of elegantly coloured true Planariæ, inhabiting the dry forest! The false relation they bear to Snails is the most extraordinary thing of the kind I have ever seen.— In the same genus (or more truly family) some of the marine species possess an organization so marvellous.—that I can scarcely credit my eyesight.— Every one has heard of the dislocoured streaks of water in the Equatorial regions.— One I examined was owing to the presence of such minute Oscillaria that in each square inch of surface there must have been at least one hundred thousand present.— After this I had better be silent.— for you will think me a Baron Munchausen amongst Naturalists.— Most assuredly I might collect a far greater number of specimens of Invertebrate animals if I took less time over each: But I have come to the conclusion, that 2 animals with their original colour & shape noted down, will be more valuable to Naturalists than 6 with only dates & place.— I hope you will send me your criticisms about my collection; & it will be my endeavour that nothing you say shall be lost on me.— ...

At this present minute we are at anchor in the mouth of the river: & such a strange scene as it is.— Every thing is in flames,—the sky with lightning,—the water with luminous particles, & even the very masts are pointed with a blue flame.— I expect great interest in scouring over the plains of M Video, yet I look back with regret to the Tropics, that magic line to all Naturalists.— The delight of sitting on a decaying trunk amidst the quiet gloom of the forest is unspeakable & never to be forgotten.— How often have I then wished for you.—

when I see a Banana, I well recollect admiring them with you in Cambridge.— little did I then think how soon I should eat their fruit.—

August 15th. . . . We have been here (Monte Video) for some time; but owing to bad weather & continual fighting on shore have scarcely ever been able to walk in the country.— I have collected during the last month nothing.— But to day I have been out & returned like Noahs ark.—with animals of all sorts.— I have to day to my astonishment found 2 *Planariæ* living under dry stones. Ask L Jenyns if he has ever heard of this fact. I also found a most curious snail & Spiders, beetles, snakes, scorpions ad libitum And to conclude shot a Cavia weighing a cwt:— On Friday we sail for the Rio Negro, & then will commence our real wild work.— I look forward with dread to the wet stormy regions of the South.— But after so much pleasure I must put up with some sea-sickness & misery.—

Remember me most kindly to every body & believe me, my dear Henslow, Yours affectionately | Chas. Darwin . . .

To J. S. Henslow [*c*. 26 October –] 24 November [1832]

Monte Video [Buenos Ayres]

My dear Henslow,

We arrived here on the 24th of Octob: after our first cruize on the coast of Patagonia: North of the Rio Negro we fell in with some little Schooners employed in sealing; to save the loss of time in surveying the intricate mass of banks, Capt: FitzRoy has hired two of them & has put officers in them.— It took us nearly a month fitting them out; as soon as this was finished we came back here, & are now preparing for a long cruize to the South.— I expect to find the wild mountainous country of Terra del. very interesting; & after the coast of Patagonia I shall thoroughily enjoy it.— I had hoped for the credit of dame Nature, no such country as this last existed; in sad reality we coasted along 240 miles of sand hillocks; I never knew before, what a horrid ugly object a sand hillock is:— The famed country of the Rio Plata in my opinion is not much better; an enormous brackish river bounded by an interminable green plain, is enough to make any naturalist groan. So hurrah for Cape Horn & the land of storms.—

Now that I have had my growl out, which is a priviledge *sailors* take on all occasions, I will turn the tables & give an account of my doings in Nat: History.— I must have one more growl, by ill luck the French government has sent one of its Collectors[2] to the Rio Negro.—where he has been working for the last six month, & is now gone round the Horn.— So that I am very selfishly afraid he will get the cream of all the good things, before me.— As I have nobody to talk to about my luck & ill luck in collecting; I am determined to vent it all upon you.— I have been very lucky with fossil bones; I have fragments of at least 6 distinct animals; as many of them are teeth I trust, *shattered & rolled* as they have been, they will be recognised. I have paid *all the attention*, I am *capable* of, to their geological site, but of course it is too long a story for here.— 1st. the Tarsi & Metatarsi very perfect of a Cavia: 2nd the upper jaw & head of some very large animal, with 4 square hollow molars.—& the head greatly produced in front.— I at first thought it belonged either to the Megalonyx or Megatherium.— In confirmation, of this, in the same formation I found a large surface of the osseous polygonal plates, which "late observations" (what are they?) show belong to the Megatherium.— Immediately I saw them I thought they must belong to an enormous Armadillo, living species of which genus are so abundant here: 3^{d} The lower jaw of some large animal, which from the molar teeth, I should think belonged to the Edentata: 4th. some large molar teeth, which in some respects would seem to belong to an enormous Rodentia; 5th, also some smaller teeth belonging to the same order: &c &c.— If it interests you sufficiently to unpack them, I shall be *very curious* to hear something about them:— *Care must be taken*, in this case, not to confuse the tallies.— They are mingled with marine shells, which appear to me identical with what now exist.— But since they were deposited in their beds, several geological changes have taken place in the country.—

So much for the dead & now for the living.— there is a poor specimen of a bird, which to my unornithological eyes, appears to be a happy mixture of a lark pidgeon & snipe ... I suppose it will turn out to be some well-know bird although it has quite baffled me.— I have taken some interesting amphibia; a fine Bipes; a new Trigonocephalus beautifully connecting in its habits Crotalus & Viperus: & plenty of new (as far as my knowledge goes) Saurians.— As for one little toad; I hope it may be new, that it may be Christened "diabolicus".— Milton must allude to this very individual, when he talks of "squat like [a] toad",[3] its colours are by Werner,[4] *ink black*, *Vermilion red* & *buff orange*.— It has been a splendid cruize for me in

Nat: History.— Amongst the pelagic Crustaceae, some new & curious genera.— In the Zoophites some interesting animals.— as for one Flustra, if I had not the specimen to back me up, nobody would believe in its most anomalous structure.— But as for novelty all this is nothing to a family of pelagic animals; which at first sight appear like Medusa, but are really highly organized.— I have examined them repeatedly, & certainly from their structure, it would be impossible to place them in any existing order.— Perhaps Salpa is the nearest animal; although the transparency of the body is nearly the only character they have in common.— All this may be said of another animal, although of a much simpler structure.—

I think the dried plants nearly contain all which were then Bahia Blanca flowering. All the specimens will be packed in casks—I think there will be three: (before sending this letter I will specify dates &c &c).— I am afraid you will groan or rather the floor of the Lecture room will, when the casks arrive.— Without you I should be utterly undone.— The small cask contains fish; will you open it, to see how the spirit has stood the evaporation of the Tropics.—

On board the Ship, everything goes on as well as possible, the only drawback is the fearful length of time between this & day of our return.— I do not see any limits to it: one year is nearly completed & the second will be so before we even leave the East coast of S America.— And then our voyage may be said really to have commenced.— I know not, how I shall be able to endure it.— The frequency with which I think of all the happy hours I have spent at Shrewsbury & Cambridge, is rather ominous.— I trust everything to time & fate & will feel my way as I go on:— We have been at Buenos Ayres for a week.— Novr. 24th.— It is a fine large city; but such a country; everything is mud; You can go no where, you can do nothing for mud.— In the city I obtained much information about the banks of the Uruguay.— I hear of Limestone with shells, & beds of shells in every direction.— I hope, when we winter in the Plata to have a most interesting Geological excursion in that country.— I purchased fragments (No^{rs}: 837 & 8) of some enormous bones; which I was assured belonged to the former *giants*!!— I also procured some seeds.— I do not know whether they are worth your accepting; if you think so, I will get some more:— They are in the box: I have sent to you by the Duke of York Packet, commanded by Lieu: Snell to Falmouth.— two large casks, containing fossil bones.—a small cask with fish, & a box containing skins, spirit bottle &c & pill-boxes with beetles.— Would you be kind enough to open these latter, as they are apt to bec⟨ome⟩ mouldy.— With the exceptions of the bones, the rest of my collection looks very scanty. Recollect how great a proportion of time is spent at sea. I am always anxious to hear in what state my things come & any criticisms about quantity or kind of specimens.— In the smaller cask is part of a large head, the anterior portions of which are in the other large ones.— The packet has arrived & I am in a great bustle: You will not hear from me for some months:

Till then believe me, my dear Henslow, Yours very truly obliged, Chas Darwin.—

Remember me most kindly to M^{rs}. Henslow.—

To J. S. Henslow 11 April 1833

April 11th.— 1833

My dear Henslow

We are now running up from the Falkland Islands to the Rio Negro (or Colorado).— The Beagle will proceed to M: Video; but if it can be managed I intend staying at the former place.— It is now some months since we have been at a civilized port, nearly all this time has been spent in the most Southern part of Tierra del Fuego.— It is a detestable place, gales succeed gales with such short intervals, that it is difficult to do anything.— We were 23 days off Cape Horn, & could by no means get to the Westward.— The last & finale gale, before we gave up the attempt was unusually severe. A sea stove one of the boats & there was so much water on the decks, that every place was afloat; nearly all the paper for drying plants is spoiled & half of this cruizes collection.— We at last run in to harbor & in the boats got to the West by the inland channels.— As I was one of this party, I was very glad of it: with two boats we went about 300 miles, & thus I had an excellent opportunity of geologising & seeing much of the Savages.— The Fuegians are in a more miserable state of barbarism, than I had expected ever to have seen a human being.— In this inclement country, they are absolutely naked, & their temporary houses are like what children make in summer, with boughs of trees.— I do not think any spectacle can be more interesting, than the first sight of Man in his primitive wildness.— It is an interest, which cannot well be imagined, untill it is experienced. I shall never forget, when entering Good Success Bay, the yell with which a party received us. They were seated on a rocky point, surrounded by the dark forest of beech; as they threw their arms wildly round their heads & their long hair streaming they seemed the troubled spirits of another world.— The climate in some respects, is a curious mixture of severity & mildness; as far as regards the animal kingdom the former character prevails; I have in consequence, not added much to my collections.— The geology of this part of Tierra del was, as indeed every place is, to me very interesting.— the country is non-fossiliferous & a common place succession of granitic rocks & Slates: attempting to make out the relation of cleavage, strata &c &c was my chief amusement.— The mineralogy however of some of the rocks, will I think be curious, from their resemblance to those of Volcanic origin.

In Zoology, during the whole cruize, I have done little; the Southern ocean is nearly as sterile as the continent it washes.— Crustaceæ have afforded me most work: it is an order most imperfectly known: I found a Zoëa, of most curious form, its body being only $\frac{1}{6}^{th}$ the length of the two spears.— I am convinced from its structure & other reasons it is a young Erichthus!— I must mention part of the structure of a Decapod, it so very anomalous: the last pair of legs are small & dorsal, but instead of being terminated by a claw, as in all others, it has three curved bristle-like appendages, these are finely serrated & furnished with cups, somewhat resembling those of the Cephalopods.— The animal being pelagic it is a beautiful structure to enable it to hold on to light floating objects.— I have found out something about the propagation of that ambiguous tribe, the Corallinas.— And this makes up nearly the poor catalogue of rarities during this cruize. After leaving Tierra del we sailed to the Falklan⟨ds.⟩ . . .

On our arrival at the Falklands everyone was much surprised to find the English flag hoisted. This our new island, is but a desolate looking spot yet must eventually be of great importance to shipping.— I had here the high good fortune, to find amongst most primitive looking rocks, a bed of micaceous sandstone, abounding with Terebratula & its subgenera & Entrochitus. As this is so remote a locality from Europe I think the comparison of these impressions, with those of the oldest fossiliferous rocks of Europe will be preeminently interesting. Of course there are only models & casts; but many of these are very perfect. I hope sufficiently so to identify species.— As I consider myself your pupil, nothing gives me more pleasure, than telling you my good luck.— I am very impatient to hear from you. When I am sea-sick & miserable, ⟨i⟩t is one of my highest consolations, to picture the future, ⟨w⟩hen we again shall be pacing together the roads round Cambridge. That day is a weary long way off: we have another cruize to make to Tierra del. next summer, & then our voyage round the world will really commence. Capt. FitzRoy has purchased a large Schooner of 170 tuns. In many respects it will be a great advantage having a consort: perhaps it may somewhat shorten our cruize: which I most cordially hope it may: I trust however that the Corall reefs & various animals of the Pacific may keep up my resolution.—

Remember me most kindly to M^{rs}. Henslow & all other friends; I am a true lover of Alma Mater, & all its inhabitants. Believe me My dear Henslow | Your affectionate & most obliged friend | Charles Darwin . . .

I am convinced from talking to the finder, that the Megatherium, sent to Geol: Soc: belongs to same formation which those bones I sent home do & that it was wa⟨she⟩d into the River from the cliffs which compose the banks: Professor Sedgwick might like to know this: & tell him I have never ceased being thankful for that short tour in Wales

To J. S. Henslow   March 1834

E. Falkland Is^{d}.

March— 1834

My dear Henslow

Upon our arrival at this place I was delighted at receiving your letter dated Aug. 31.— Nothing for a long time has given me so much pleasure. Independent of this pleasure, your account of the safe arrival of my second cargo & that some of the Specimens were interesting, has been, as you may well suppose, most highly satisfactory to me.—

I am quite astonished that such miserable fragments of the Megatherium should have been worth all the trouble Mr Clift has bestowed on them. I have been alarmed by the expression cleaning all the bones, as I am afraid the printed numbers will be lost: the reason I am so anxious they should not be, is that a part were found in a gravel with recent shells, but others in a very different bed.— Now with these latter there were bones of an Agouti, a genus of animals I believe now peculiar to America & it would be curious to prove some one of the same genus coexisted with the Megatherium; such & *many other* points **entirely** depend on the numbers being carefully preserved.— My entire ignorance of comparative Anatomy makes me quite dependent on the numbers: so that you will see my geological notes will be useless without I am certain to what specimens I refer.— Since receiving these specimens, you ought to have received two others cargos, shipped from the Plata in July & November 1833.— With the latter there was a heavy box of fossil remains, which is now I suppose at Plymouth. I followed this plan from not liking to give you so much trouble: it contains another imperfect Megatherium head, & some part of the skeleton of an animal, of which I formerly sent the jaw, which had four teeth on each side in shape like this ◯ ▱ ᴏ ᴏ .— I am curious to know to what it belongs.—

Shortly before I left M: Video I bought far up in the country for two shillings a head of a Megatherium which must have been when found quite perfect.— The Gauchos however broke the teeth & lost the lower jaw, but the lower & internal parts are tolerably perfect: It is now, I hope, on the high seas in pursuit of me.— It is a most flattering encouragement to find Men, like Mr Clift, who will take such interest, in what I send home.—

I am very glad the plants give you any pleasure; I do assure you I was so ashamed of them, I had a great mind to throw them away; but if they give you any pleasure I am indeed bound, & will pledge myself to collect whenever we are in parts not often visited by Ships & Collectors.— I collected all the plants, which were in flower on the coast of Patagonia at Port Desire & St. Julian; also on the Eastern parts of Tierra del Fuego, where the climate & features of T del Fuego & Patagonia are united. With them there are as many seeds, as I could find (you had better plant all ye rubbish which I send, for some of the seeds were very small).— The soil of Patagonia is *very* dry, *gravelly* & light.— in East Tierra, it is gravelly—peaty & damp.— Since leaving the R. Plata, I have had some opportunities of examining the great Southern Patagonian formation.— I have a good many shells; from the little I know of the subject it must be a Tertiary formation for some of the shells & (Corallines?) now exist in the sea.— others I believe do not.— This bed, which is chiefly characterised by a great Oyster is covered by a very curious bed of Porphyry pebbles, which I have traced for more than 700 miles.—but the most curious fact is that the whole of the East coast of South part of S. America has been elevated from the ocean, since a period during which Muscles have not lost their blue color.—

At Port St Julian I found some very perfect bones of some large animal, I fancy a Mastodon.— the bones of one hind extremity are very perfect & solid.— This is interesting as the Latitude is between 49° & 50° & the site is so far removed from the great Pampas, where bones of the narrow toothed Mastodon are so frequently found— By the way this Mastodon & the Megatherium, I have no doubt were fellow brethren in the ancient plains Relics of the Megatherium I have found at a distance of nearly 600 miles apart in a N & S. line.—

In Tierra del Fuego I have been interested in finding some sort of Ammonite (also I believe found by Capt King) in the Slate near Port Famine; on the Eastern coast there are some curious alluvial plains, by which the existence of certain quadrupeds in the islands can clearly be accounted for.— There is a sandstone, with the impression of the leaves of the common Beech tree also modern shells, &c &c.— On the surface of which table land there are, as usual, muscles with their blue color &c.— This is the *report* of my *geological section!* to you my President & Master.— I am quite charmed with Geology but like the wise animal between two bundles of hay, I do not know which to like the best, the old crystalline group of rocks or the softer & fossiliferous beds.— When puzzling about stratification &c, I feel inclined to cry a fig for your big oysters & your bigger Megatheriums.— But then when digging out some fine bones, I wonder how any man can tire his arms with hammering granite.— By the way I have not one clear idea about cleavage, stratification, lines of upheaval.— I have no books, which tell me much & what they do I cannot apply to what I see. In consequence I draw my own conclusions, & most gloriously ridiculous ones they are, I sometimes fancy I shall persuade myself there are no such things as mountains, which would be a very original discovery to make in Tierra del Fuego.— Can you throw any light into my mind, by telling me what relation cleavage & planes of deposition bear to each other?—

And now for my second *section* Zoology.— I have chiefly been employed in preparing myself for the South sea, by examining the Polypi of the smaller Corallines in these latitudes.— Many in themselves are very curious, & I think are quite undescribed, there was one appalling one, allied to a Flustra which I daresay I mentioned having found to the Northward, where the cells have a moveable organ (like a Vultures head, with a dilatable beak), fixed on the edge. But what is of more general interest is the unquestionable (as

31

it appears to me) existence of another species of ostrich, besides the Struthio Rhea.— All the Gauchos & Indians state it is the case: & I place the greatest faith in their observations.— I have head, neck, piece of skin, feathers, & legs of one. The differences are chiefly in color of feathers & scales on legs, being feathered below the knees; nidification & geographical distribution.—

So much for what I have lately done; the prospect before me is full of sunshine: fine weather, glorious scenery, the geology of the Andes; plains abounding with organic remains, (which perhaps I may have the good luck to catch in the very act of moving); and lastly an ocean & its shores abounding with life.— So that, if nothing unforeseen happens I will stick to the voyage; although, for what I can see, this may last till we return a fine set of whiteheaded old gentlemen.—

I have to thank you most cordially for sending me the Books.— I am now reading the Oxford Report.— [of the British Association for the Advancement of Science] the whole account of your proceedings is most glorious; you, remaining in England, cannot well imagine how excessively interesting I find the reports; I am sure, from my own thrilling sensations, when reading them, that they cannot fail to have an excellent effect upon all those residing in distant colonies, & who have little opportunity of seeing the Periodicals.— My hammer has flown with redoubled force on the devoted blocks; as I thought over the eloquence of the Cambridge President [Adam Sedgwick] I hit harder & harder blows. I hope, to give my arm strength for the Cordilleras, you will send me, through Capt. Beaufort, a copy of the Cambridge Report.—

I have forgotten to mention, that for some time past & for the future I will put a pencil cross on the pill-boxes containing insects; as these alone will require being kept particularly dry, it may perhaps save you some trouble.—

When this letter will go, I do not know, as this little seat of discord has lately been embroiled by a dreadful scene of murder & at present there are more prisoners, than inhabitants.— If a merchant vessel is chartered to take them to Rio I will send some specimens (especially my few plants & seeds).—

Remember me to all my Cambridge friends.— I love & treasure up every recollection of dear old Cambridge.— ...

Farewell my dear Henslow— believe my your most obliged & affectionate friend. Charles Darwin.— ...

The Voyage: South America – West Coast

Valparaiso
July 24th.— 1834

My dear Henslow

A box has just arrived, in which are two of your most kind & affectionate letters; you do not know how happy they have made me. . . . Not having heard from you untill March of this year; I really began to think my collections were so poor, that you were puzzled what to say: the case is now quite on the opposite tack; for you are **guilty** of exciting all my vain feelings to a most comfortable pitch; if hard work will atone for these thoughts I vow it shall not be spared.—

It is rather late, but I will allude to some remarks in the Jan: letter: you advise me to send home duplicates of my notes; I have been aware of the advantage of doing so; but then at sea to this day, I am invariably sick, excepting on the finest days; at which times with pelagic animals around me, I could never bring myself to the task; on shore, the most prudent person, could hardly expect such a sacrifice of time.—

My notes are becoming bulky; I have about 600 small quarto pages full; about half of this is Geology, the other imperfect descriptions of animals: with the latter I make it a rule only to describe those parts, or facts, which cannot be seen, in specimens in spirits. I keep my private Journal distinct from the above.— (NB this letter is a most untidy one, but my mind is untidy with joy; it is your *fault*, so you must take the consequence). With respect to the land Planariæ: unquestionably they are not Molluscous animals: I read your letters last night, this morning I took a little walk; by a curious coincidence I found a new white species of Planaria & a (new to me) Vaginulus (3^{d} species which I have found in S. America) of Cuv: I suppose this is the animal Leonard Jenyns alludes to.— The *true Onchidium* of **Cuv**: I likewise know.— Amongst the marine Mollusques I have seen a good many genera & at Rio found one quite new one.— With respect to

33

the December letter, I am very glad to hear, the four casks arrived safe; since which time you will have received another cargo, with the bird skins, about which you did not understand me.— Have any of the B. Ayrean seeds produced plants?—

From the Falklands, I acknowledged a box & letter from you; with the letter were a few seeds from Patagonia.— At present, I have specimens enough to make a heavy cargo, but shall wait as much longer as possible, because opportunities are not now so good as before.— I have just got scent of some fossil bones of a **Mammoth**!, what they may be, I do not know, but if gold or galloping will get them, they shall be mine. ...

After leaving the Falklands, we proceeded to the R. S. Cruz; followed up the river till within 20 miles of the Cordilleras: Unfortunately want of provisions compelled us to return. This expedition was most important to me, as it was a transverse section of the great Patagonian formation.— I conjecture (an accurate examination of fossils may possibly determine the point) that the main bed is somewhere about the Meiocene period, (using Mr Lyell's expression) I judge from what I have seen of the present shells of Patagonia.— This bed contains an *enormous* field of Lava.— This is of some interest, as being a rude approximation to the age of the Volcanic part of the great range of the Andes.— Long before this it existed as a Slate & *Porphyritic* line of hills.— I have collected tolerable quantity of information respecting the period, (even numbers) & forms of elevations of these plains. I think these will be interesting to Mr Lyell.— I had deferred reading his third volume till my return, you may guess how much pleasure it gave me; some of his wood-cuts came so exactly into play, that I have only to refer to them, instead of redrawing similar ones.— I had my Barometer with me; I only wish I had used it more in these plains.—

The valley of S. Cruz appears to me a very curious one, at first it quite baffled me.— I believe I can show good reasons for supposing it to have been once a **Northern** St^{ts}. like that of *Magellan*.— When I return to England, you will have some hard work in winnowing my Geology; what little I know, I have learnt in such a curious fashion, that I often feel very doubtful about the number of grains: Whatever number, they may turn out, I have enjoyed extreme pleasure in collecting them.—

In T. del Fuego I collected & examined some Corallines: I have observed one fact which quite startled me.— it is, that in the genus

Sertularia, (taken in its most restricted form as by Lamouroux)[1] & in 2 species which, excluding comparative expressions, I should find much difficulty in describing as different—the Polypi quite & essentially differed, in all their most important & evident parts of structure.— I have already seen enough to be convinced that the present families of Corallines, as arranged by Lamarck, Cuvier &c are highly artificial.— It appears they are in the same state, which shells were when Linnæus left them for Cuvier to rearrange.—

I do so wish I was a better hand at dissecting: I find I can do very little in the minute parts of structure; I am forced to take a very rough examination as a type for different classes of structure. ...

The Beagle left the St[s] of Magellan in the middle of winter; she found her road out by a wild unfrequented channel; well might Sir J. Narborough call the West coast South Desolation "because it is so desolate a land to behold".—[2] We were driven into Chiloe, by some very bad weather.— an Englishman gave me 3 specimens of that very fine *Lucanoidal* insect, which is described Camb: Phil. Trans: 2 males & one female.— I find Chiloe is composed of Lava & recent deposits.— the Lavas are curious from abounding or rather being in parts composed of Pitchstone.— If we go to Chiloe in the summer I shall reap an Entomological harvest.— I suppose the Botany both there & in Chili is well known.—

I forgot to state, that in the four cargoes of specimens there have been sent 3 square boxes, each containing four glass bottles.— I mention this in case they should be stowed beneath geological specimens, & thus escape your notice perhaps some spirit may be wanted in them.— If a box arrives from B. Ayres, with Megatherium head & other *unnumbered* specimens: be kind enough to tell me; I have strong fears for its safety.—

We arrived here the day before yesterday; the views of the distant mountains are most sublime & the climate delightful; after our long cruize in the damp gloomy climates of the South, to breathe a clear, dry air, & feel honest warm sunshine, & eat good fresh roast beef must be the summum bonum of human life.— I do not like the looks of the rocks, half so much as the beef, there is too much of those rather insipid ingredients Mica, quartz & Feldspar. ... I have sent you in this letter a sad dose of egotism.—but recollect I look up to you as my father in Natural History, & a son may talk about himself, to his father.— In your paternal capacity, as pro-proctor what a great deal of trouble you appear to have had.— ...

October 28th.— This letter has been lying in my port-folio ever since July: I did not send it away, because I did not think it worth the postage: it shall now go with a box of specimens: shortly after arriving here, I set out on a geological excursion, & had a very pleasant ramble about the base of the Andes.— The whole country appears composed of breccias, (& I imagine Slates) which universally have been modified, & oftentimes completely altered by the action of fire; the varieties of porphyry thus produced is endless, but no where have I yet met with rocks which have flowed in a stream; dykes of greenstone are very numerous: Modern Volcanic action is entirely shut up in the very central parts (which cannot now be reached on account of the snow) of the Cordilleras.— To the South of the R. Maypo I examined the Tertiary plains already partially described by M. Gay.³ The fossil shells, appear to me, to be far more different from the recent ones, than in the great Patagonian formation; it will be curious if an Eocene & Meiocene (Recent there is abundance of) could be proved to exist in S. America as well as in Europe.— I have been much interested by finding abundance of recent shells at an elevation of 1300 feet; the country in many places is scattered over with shells, but these are *all littoral* ones. So that I suppose the 1300 feet elevation *must* be owing to a succession of small elevations such as in 1822. With these certain proofs of the recent residence of the ocean over all the lower parts of Chili; the outline of every view & the form of each valley possesses a high interest. Has the action of running water or the sea formed this deep ravine? Was a question which often arose in my mind, & generally was answered by finding a bed of recent shells at the bottom.— I have not sufficient arguments, but I do not believe that more than a small fraction of the height of the Andes has been formed within the Tertiary period.—

The conclusion of my excursion was very unfortunate, I became unwell & could hardly reach this place, I have been in bed for the last month, but am now rapidly getting well. I had hoped during this time to have made a good collection of insects &c but it has been impossible. I regret the less, because Chili fairly swarms with Collectors; there are more Naturalists in the country, than Carpenters or Shoemaker or any other honest trade.—

In my letter from the Falkland Is^d. I said I had fears about a box with a Megatherium. I have since heard from B. Ayres, that it went to Liverpool by the Brig Basingwaithe.— If you have not received it—it is, I think, worth taking some trouble about. In October two casks & a jar were sent by H.M.S. Samarang viâ Portsmouth I have no doubt you have received them. With this letter, I send a good many bird skins; in the same box with them, there is a paper parcel, containing pill boxes with insects: the other pill-boxes require no particular care: You will see in two of these boxes, some dried terrestrial Planariæ, the only method I have found of preserving them (they are exceedingly brittle) By examining the white species I understand some little of the internal structure.— There are two small parcels of seeds.— There are some plants, which I hope may interest you, or at least those from Patagonia, where I collected every one in flower:— There is a bottle, clumsily, but I think securely corked, containing water & *gaz* from the hot Baths of Cauquenes, seated at foot of Andes & long celebrated for medicinal properties.— I took pains in filling & securing both water & gaz.— If you can find any one who likes to analyze them; I should think it would be worth the trouble.— I have not time at present to copy my few observations about the locality &c &c of these Springs.— Will you tell me, how the Arachnidæ, which I have sent home, for instance those from Rio appear to be preserved.— I have doubts whether it is worth while collecting them.—

We sail the day after tomorrow: our plans are at last limited & definite: I am delighted to say we have bid an eternal adieu to T. del Fuego.— The Beagle will not proceed further South than C. Tres Montes. From which point we survey to the North. The Chonos archipelago is delightfully unknown; fine deep inlets running into the Cordilleras, where we can steer by the light of a Volcano.— I do not know, which part of the voyage, now offers the most attractions.— This is a shamefully untidy letter, but you must forgive me & believe me | My dear Henslow | Yours most truly obliged | Charles Darwin Nov.^b 7th.—

To Caroline Darwin 13 October 1834

Valparaiso.
October 13th. 1834.

My dear Caroline

I have been unwell & in bed for the last fortnight, & am now only able to sit up for a short time. As I want occupation I will try & fill this letter.— Returning from my excursion into the country I staid a few days at some Goldmines & whilst there I drank some Chichi a very weak, sour new made wine, this half poisoned me, I staid till I thought I was well; but my first days ride, which was a long one again disordered my stomach, & afterwards I could not get well; I quite lost my appetite & became very weak. I had a long distance to travel & I suffered very much; at last I arrived here quite exhausted. But Bynoe with a good deal of Calomel & rest has nearly put me right again & I am now only a little feeble.— I consider myself very lucky in having reached this place, without having tried it, I should have thought it not possible; a man has a great deal more strength in him, when he is unwell, than he is aware of. If it had not been for this accident, my ride would have been very pleasant. I made a circuit, taking in St Iago. I set out by the valley of Aconcagua I had some capital scrambling about the mountains. I slept two nights near the summit of the Bell of Quillota. This is the highest mountain out of the chain of the Andes, being 4700 ft high. The view was very interesting, as it afforded a complete map of the Cordilleras & Chili.— From here I paid a visit to a Cornish miner who is working some mines in a ravine in the very Andes. I throughily enjoyed rambling about, hammer in hand, the bases of these great giants, as independently as I would the mountains in Wales. I reached the Snow but found it quite impossible to penetrate any higher.— I now struck down to the South, to St Iago the gay Capital of Chili. ... St Iago is built on a plain; the basin of a former inland sea; the perfect levelness of this plain is contrasted in a strange & picturesque manner with great, snow topped mountains, which surround it.— From St Iago I proceeded to S. Fernando about 40 leagues to the South.— Every one in the city talked so much about the robbers & murderers, I was persuaded to take another man with me, this added very much to the expense; & now I do not think it was necessary. Altogether it has been the most expensive excursion, I ever made, & in return I have seen scarcely enough of the Geology to repay it.— I was however lucky in getting a good many fossil shells from the modern formation of Chili.—

On my road to S. Fernando, I had some more hammering at the Andes, as I staid a few days at the hot springs of Cauquenes, situated in one of the valleys.— From S. Fernando I cut across the country to the coast & then returned, as I have said very miserable to Corfields house here at Valparaiso. You will be sorry to hear, the Schooner, the Adventure is sold; the Captain received no sort of encouragement from the Admiralty & he found the expense ⟨of⟩ so large a vessel so immense he determined at once to ⟨give⟩ her up.— We are now in the same state as when we left England with Wickham for 1st Lieut, which part of the business anyhow is a good job.— we shall all be very badly off for room; & I shall have trouble enough with stowing my collections. It is in every point of view a grievous affair in our little world; a sad tumbling down for some of the officers, from 1st. Lieut of the Schooner to the miserable midshipmans birth.—& many similar degradations.— It is necessary also to leave our little painter, Martens, to wander about ye world.— Thank Heavens, however, the Captain positively asserts that this change shall not prolong the voyage.—that in less than 2 years we shall be at New S. Wales.—

I find being sick at stomach inclines one also to be home-sick. In about a fortnight the Beagle proceeds down the coast, touches at Concepcion & Valdivia & sets to work behind Chiloe. I suspect we shall pay T del Fuego another visit; but of this good Lord deliver us: it is kept very secret, lest the men should desert; every one so hates the confounded country. Our voyage sounded much more delightful in the instructions, than it really is; in fact it is a survey of S. America, & return by the C. of Good Hope instead of C. Horn. We shall see nothing of any country, excepting S. America. But I ought not to grumble, for the voyage is for this very reason, I believe, much better for my pursuits, although not nearly so agreeable as a tour.— I will write again before sailing. I am however at present deeply in debt with letters. I received shortly since a very kind long one from Mr Owen, which I will shortly answer.— Letter writing is a task, which I throughly dislike.— I do not mean writing to home: but to any body else, for really after such interval I have nothing to tell but my own history, & that is very tedious.—

I have picked up one very odd correspondent, it is Mr Fox the Minister at Rio. (it is the Mr Fox, who in one of Lord Byrons letters is said to be so altered after an illness that his *oldest Creditors* would not know him) ...

We are all here in great anxiety to hear some political news. A Ship sailed from Liverpool just after Ld Greys resignation & we cannot guess who will succeed him.—[4]

Give my best love to my Father & all of you & Believe me my very dear Caroline | Yours affectionately | Charles Darwin.—

To Catherine Darwin 8 November 1834

Valparaiso.

November 8th. 1834

My dear Catherine

My last letter was rather a gloomy one, for I was not very well when I wrote it— Now everything is as bright as sunshine. I am quite well again after being a second time in bed for a fortnight. Capt FitzRoy very generously has delayed the Ship 10 days on my account & without at the time telling me for what reason.— We have had some strange proceedings on board the Beagle, but which have ended most capitally for all hands.— Capt FitzRoy has for the last two months, been working **extremely** hard & at same time constantly annoyed by interruptions from officers of other ships: the selling the Schooner & its consequences were very vexatious: the cold manner the Admiralty (solely I believe because he is a Tory) have treated him, & a thousand other &c &c has made him very thin & unwell, This was accompanied by a morbid depression of spirits, & a loss of all decision & resolution

The Captain was afraid that his mind was becoming deranged (being aware of his heredetary predisposition). all that Bynoe could say, that it was merely the effect of bodily health & exhaustion after such application, would not do; he invalided & Wickham was appointed to the command. By the instructions Wickham could only finish the survey of the Southern part & would then have been obliged to return direct to England.— The grief on board the Beagle about the Captains decision was universal & deeply felt.— One great source of his annoyment, was the feeling it impossible to fulfil the whole instructions; from his state of mind, it never occurred to him, that the very instructions order him to do as much of West coast, as *he has time* for & then proceed across the Pacific. Wickham (very disinterestedly, giving up his own promotion) urged this most strongly, stating that when he took the command, nothing should induce him to go to T. del Fuego again; & then asked the Captain, what would be gained by his resignation Why not do the more useful part & return, as commanded by the Pacific. The Captain, at last, to every ones joy consented & the resignation was withdrawn.—

Hurra Hurra it is fixed the Beagle shall not go one mile South of C. Tres Montes (about 200 miles South of Chiloe) & from that point to Valparaiso will be finished in about five months.— We shall examine the Chonos archipelago, entirely unknown & the curious inland sea behind Chiloe.— For me it is glorious C. T. Montes is the most Southern point where there is much geological interest, as there the modern beds end.— The Captain then talks of crossing the Pacific; but I think we shall persuade him to finish the coast of Peru: where the climate is delightful, the country hideously sterile but abounding with the highest interest to a Geologist. For the first time since leaving England I now see a clear & not so distant prospect of returning to you all: crossing the Pacific & from Sydney home will not take much time.—

As soon as the Captain invalided, I at once determined to leave the Beagle; but it was quite absurd, what a revolution in five minutes was effected in all my feelings. I have long been grieved & most sorry at the interminable length of the voyage (although I never would have quitted it).—but the minute it was all over, I could not make up my mind to return, I could not give up all the geological castles in the air, which I had been building for the last two years.— One whole night I tried to think over the pleasure of seeing Shrewsbury again, but the barren, plains of Peru gained the day. I made the following scheme. (I know you will abuse me, & perhaps if I had put it in execution my Father would have sent a mandamus after me), it was to examine the Cordilleras of Chili during this summer & in the winter go from Port to Port on the coast of Peru to Lima returning this time next year to Valparaiso, cross the Cordilleras to B. Ayres & take ship to England.— Would this not have been a fine excursion & in 16 months I should have been with you all. To have endured T. del F. & not seen the Pacific would have been miserable: As things are at present, they are perfect; the intended completion of *small* parts of the survey of S.W coast would have possessed no interest & the Coast is in fact frightfully dangerous, & the climate worse than about C. Horn.— When we are once at sea, I am sure the Captain will be all right again; he has already regained his cool inflexible manner, which he had quite lost.—

I go on board tomorrow; I have been for the last six weeks in Corfields house. You cannot imagine what a kind friend I have found him.— He is universally liked & respected by the Natives & Foreigners.— Several Chileno Signoritas are very obligingly anxious to become the Signoras of this house.— Tell my Father, I have kept my promise of being extravagant in Chili. I have drawn a bill of 100£ (Had it not better be notified to Mr Robarts & Co?). 50£ goes to the Captain for ensuing year & 30 I take to sea for the small ports; so that bonâ fide I have not spent 180 during these last four months.— I hope not to draw another bill for 6 months. All the foregoing particulars were only settled yesterday: it has done me more good that a pint of Medicin; & I have not been so happy for the last year.— If it had not been for my illness, these four months in Chili, would have been very pleasant: I have had ill luck however in only one little earthquake having happened.— I was lying in bed, when there was a party at dinner, in the house; on a sudden I heard such a hubbub in the dining room; without a word being spoken, it was devil take the hind most who should get out first: at the same moment I felt my bed *slightly* vibrate in a lateral direction. The party were old stagers & heard, the noise, which always precedes a shock; & no old Stager looks at an earthquake with philosophical eyes. ...

My dear Catherine. Your affectionately | Chas. Darwin ...

To Susan Darwin 23 April 1835

Valparaiso
April 23^d.— 1835

My dear Susan

I received a few days since your letter of November: the three letters, which I before mentioned are yet missing: but I do not doubt they will come to life.— I returned a week ago from my excursion across the Andes to Mendoza. Since leaving England I have never made so successful a journey: it has however been very expensive: I am sure my Father would not regret it, if he could know how deeply I have enjoyed it.— it was something more than enjoyment: I cannot express the delight, which I felt at such a famous winding up of all my geology in S.— America.— I literally could hardly sleep at nights for thinking over my days work.— The scenery was so new & so majestic: every thing at an elevation of 12000 ft. bears so different an aspect, from that in a lower country.— I have seen many views more beautiful but none with so strongly marked a character. To a geologist also there are such manifest proofs of excessive violence, the strata of the highest pinnacles are tossed about like the crust of a broken pie. I crossed by the Portillo pass, which at this time of year is apt to be dangerous, so could not afford to delay there; after staying a day in the stupid town of Mendoza I began my return by Uspallata, which I did very leisurely.— My whole trip only took up 22 days.— I travelled with, for me, uncommon comfort, as I carried a *bed*!: my party consisted of two Peons & 10 mules, two of which were with baggage or rather food, in case of being snowed up.— Every thing however favoured me, not even a speck of this years Snow had fallen on the road.—

I do not suppose, any of you can be much interested in Geological details, but I will just mention my principal, results: beside understanding to a certain extent, the description & manner of the force, which has elevated this great line of mountains, I can clearly demonstrate, that one part of the double line is of a age long posterior to the other. In the more ancient line, which is the true chain of the Andes.—I can describe the sort & order of the rocks which compose it. These are chiefly remarkable by containing a bed of Gypsum nearly 2000 ft thick: a quantity of this substance I should think unparalleled in the world. What is of much greater consequence, I have procured fossil shells (from an elevation of 12000 ft) I think an examination of these will give an approximate age to these mountains as compared to the Strata of Europe: In the other line of the Cordilleras there is a strong presumption (in my own mind conviction) that the enormous mass of mountains, the peaks of which rise to 13 & 14000 ft are so very modern as to be contemporaneous with the plains of Patagonia (or about with *upper* strata of Isle of Wight): If this result shall be considered as proved it is a very important fact in the theory of the formation of the world.— Because if such wonderful changes have taken place so recently in the crust of the globe, there can be no reason for supposing former epochs of excessive violence.— These modern strata are very remarkable by being threaded with metallic veins of Silver, Gold, Copper &c: hitherto, these have been considered as appertaining to older formations. In these same beds (& close to a Gold mine) I found a clump of petrified trees, standing upright, with the layers of fine Sandstone deposited round them, bearing the impression of their bark. These trees are covered by other Sandstones & streams of Lava to the thickness of several thousand feet. These rocks have been deposited beneath water, yet it is clear the spot where the trees grew, must once have been above the level of the sea, so that it is certain the land must have been depressed by at least as many thousand feet, as the superincumbent subaqueous deposits are thick.— But I am afraid you will tell me, I am prosy with my geological descriptions & theories.—

You are aware, that plants of Arctic regions are frequently found in lower latitudes, at an elevation which produces an equal degree of cold.— I noticed a rather curious illustration of this law in finding on the patches of perpetual Snow, the famous Red Snow of the Northern Navigators.— I am going to send to Henslow, a description of this little Lichen, for him, if he thinks it worth while to publish in some of the Periodicals.—

I am getting ready my last Cargo of Specimens to send to England; This last trip has added half a mule's load; for without plenty of proof I do not expect a word of what I have above written to be believed.— I arrived at this place a week since, & am as before living with Corfield. I have found him as kind & good-natured a friend as he is a good man.— I staid also a week in St Iago, to rest after the Cordilleras, of which I stood in need & lived in the house of Mr Caldcleugh (the author of some bad travels in S. America): he is a very pleasant person & took an infinite degree of trouble for me.— It is quite surprising how kind & hospitable I have found all the English merchants.— Do mention to Mr Corfield of Pitchford, under what obligations I lie to his son.— Amongst the various pieces of news, of which your letter is full, I am indeed very sorry to hear of poor Col. Leighton's death. I can well believe how much he is regretted. It is a bitter reflection, when I think what changes will have taken place before I return. I pray to Heaven I may return to see all of you. ...

The Beagle after leaving me here, returned to Concepcion: Capt Fitz Roy has investigated with admirable precision the relative level of land & Water, since the great Earthquake.— The rise is unequal & parts of the coast are now settling down again, probably at each little trembling which yet continue.— The Is^d of S. Maria has been elevated 10 feet: Capt Fitz Roy found a bed of Muscles with putrid fish that many feet above high water mark.— The Beagle passed this port yesterday. I hired a boat & pulled out to her. The Capt is very well; I was the first to communicate to him his promotion. He is fully determined, nothing shall induce him to delay the voyage a month: if time is lost in one place, something else shall be sacrificed.— Our voyage now will solely consist in carrying a chain of longitudes between important positions.

My holidays extend till the middle of July: so that I have 10 weeks before me, & the Beagle will pick me up at any Port I choose. The day after tomorrow I start for Coquimbo. I have three horses & a baggage Mule, & a Peon whom I can trust, having now accompanied me on every excursion. The people moreover to the *North*, have a capital character for honesty, ie they are not cutthroats. The weather there also will not be hot & it never rains.— I shall extend my journey to Copiapo.— it is a great distance, but I feel certain I shall be most amply repaid. Everything which can interest a Geologist, is found in those districts, Mines of Rock-Salt, Gypsum, Saltpetre, Sulphur; the rocks threaded with metallic veins: old sea-beachs;—curious formed valleys; petrified shells, Volcanoes & strange scenery. The country geologically is entirely unknown (as indeed is the whole of South S. America), & I thus shall see the whole of Chili from the Desert of Atacama to the extreme point of Chiloe. All this is very brilliant, but now comes the black & dismal part of the Prospect.—that horrid phantom, money. The country where I am going to is very thinly inhabited & it will be impossible to draw bills.— I am therefore obliged to draw the money here & transmit it there.— Moreover it is necessary to be prepared for accidents: horses stolen.—I robbed.—Peon sick, a pretty state I should be 400 or 500 miles from where I could command money.— In short, I have drawn a bill for £100: : o: : o, & this so shortly after having spent 60 in crossing the Andes. In September we leave the coast of America: & my Father will believe, that I *will* not draw money in crossing the Pacific, because I *can* not.— I verily believe I could spend money in the very moon.— My travelling expences are nothing; but when I reach a point, as Coquimbo, whilst my horses are resting, I hear of something very wonderful 100 miles off. A muleteer offers to take me for so many dollars, & I cannot or rather never have resisted the Temptation.—

My Fathers patience must be exhausted: it will be patience smiling at his son, instead of at grief. I write about it as a good joke, but upon my honor I do not consider it so.— Corfield cashes the bill & sends it to his Father, who will bring it to the old Bank, where I suppose it can be transacted.—

I received a long & affectionate letter from Fox: he alludes to a letter which I have never received. I shall write to him from Lima; at present I have my hands full.— How strange it sounds to hear him talk of "his dear little wife". Thank providence he did not marry the simple charming Bessy.— I shall be very curious to hear a verdict concerning the merits of the Lady.— How the world goes round; Eyton married. I hope he will teach his wife to sit upright.— I have written to him: I am sure he deserves to be happy. . . .

Your account of Erasmus' (does Erasmus live with the Hensleigh's for the last year their names have never in any letter been separated) visit to Cambridge has made me long to be back there. I cannot fancy anything more delightful than his Sunday round, of King's, Trinity & those talking giants, Whewell & Sedgwick: I hope your musical tastes continue in due force. I shall be ravenous for the Piano-forte. Do you recollect, poor old Granny, how I used to torment your quiet soul every evening?— I have not quite determined whether I will sleep at the Lion, the first night, when I arrive per Wonder[5] or disturb you all in the dead of the night, everything short of that is absolutely planned.— Everything about Shrewsbury is growing in my mind bigger & more beautiful; I am certain the Acacia & Copper Beech are two superb trees: I shall know every bush, & I will trouble you young ladies, when each of you cut down your tree to spare a few. As for the view behind the house I have seen nothing like it. It is the same with North Wales. Snowden to my mind, looks much higher & much more beautiful than any peak in the Cordilleras. So you will say, with my benighted faculties, it is time to return, & so it is, & I long to be with you— Whatever the trees are, I know what I shall find all you.— I am writing nonsense—so Farewell.— My most affectionate love to all & I pray forgiveness from my Father. Yours most affectionately | Charles Darwin ...

To J. S. Henslow 12 [August] 1835

Lima

July[6] 12th.— 1835

My dear Henslow

This is the last letter, which I shall ever write to you from the shores of America.— and for this reason I send it.— In a few days time the Beagle will sail for the Galapagos Isds.— I look forward with joy & interest to this, both as being somewhat nearer to England, & for the sake of having a good look at an active Volcano.— Although we have seen Lava in abundance, I have never yet beheld the Crater.— I sent by H.M.S. Conway two large boxes of Specimens. The Conway sailed the latter end of June.— With them were letters for you.— Since that time I have travelled by land from Valparaiso to Copiapò & seen something more of the Cordilleras.— Some of my Geological views have been subsequently to the last letter altered.— I believe the upper mass of strata are not so very modern as I supposed.— This last journey has explained to me much of the ancient history of the Cordilleras.— I feel sure they formerly consisted of a chain of Volcanoes from which enormous streams of Lava were poured forth at the bottom of the sea.— These alternate with sedimentary beds to a vast thickness: at a subsequent period these Volcanoes must have formed Islands, from which have been produced strata several thousand feet thick of coarse Conglomerate.— These Islands were covered with fine trees; in the Conglomerate I found one 15 feet in circumference, perfectly silicified to the very centre.— The alternations of compact crystalline rocks (I cannot doubt subaqueous Lavas) & sedimentary beds, now upheaved, fractured & indurated form the main range of the Andes. The formation was produced at the time, when *Ammonites*, several Terebratulæ, Gryphites, Oysters, Pectens, Mytili &c &c lived.—

In the central parts of Chili, the structure of the lower beds are rendered very obscure by the Metamorphic action, which has rendered even the coarsest Conglomerates, porphyritic.— The Cordilleras of the Andes so worthy of admiration from the grandeur of their dimensions, to rise in dignity when it is considered that since the period of Ammonites, they have formed a marked feature in the Geography of the Globe.— The geology of these Mountains pleased me in one respect; when reading Lyell, it had always struck me that if the crust of the world goes on changing in a Circle, there ought to be somewhere found formations which having the *age* of the great Europæan secondary beds, should possess the *structure* of Tertiary rocks, or those formed amidst Islands & in limited Basins. Now the alternations of Lava & coarse sediment, which form the upper parts of the Andes, correspond exactly to what would accumulate under such circumstances. In consequence of this I can only very *roughly* separate into three divisions the varying strata (perhaps 8000 ft thick) which compose these mountains. I am afraid you will tell me to learn my A.B.C.—to know quartz from Feldspar—before I indulge in such speculations.— I lately got hold of ⟨   ⟩ report on M. Dessalines D'Orbigny's labors in S. America. I experienced rather a debasing degree of vexation to find he has described the geology of the Pampas, & that I have had some hard riding for nothing; it was however gratifying that my conclusions are the same, as far as I can collect, with his results.— It is also capital, that the whole of Bolivia will be described. I hope to be able to connect his Geology of that country, with mine of Chili.— After leaving Copiapò, we touched at Iquique. I visited, but do not quite understand the position of the Nitrate of Soda beds.— Here in Peru, from the state of Anarchy, I can make no expedition. . . .

Believe me, dear Henslow, Yours affectionately obliged | Charles Darwin

Homeward Bound

To Caroline Darwin 27 December 1835

Bay of Islands.— New Zealand.

Decemb 27th. 1835.—

My dear Caroline,

   My last letter was written from the Galapagos,[1] since which time I have had no opportunity of sending another. A Whaling Ship is now going direct to London & I gladly take the chance of a fine rainy Sunday evening of telling you how we are getting on.— You will see we have passed the Meridian of the Antipodes & are now on the right side of the world. For the last year, I have been wishing to return & have uttered my wishes in no gentle murmurs; But now I feel inclined to keep up one steady deep growl from morning to night.— I count & recount every stage in the journey homewards & an hour lost is reckoned of more consequence, than a week formerly. There is no more Geology, but plenty of sea-sickness; hitherto the pleasures & pains have balanced each other; of the latter there is yet an abundance, but the pleasures have all moved forwards & have reached Shrewsbury some eight months before I shall.—

   If I can grumble in this style, now that I am sitting, after a very comfortable dinner of fresh pork & potatoes, quietly in my cabin, think how aimiable I must be when the Ship in a gloomy day is pitching her bows against a head Sea. Think, & pity me.— But everything is tolerable, when I recollect that this day eight months I probably shall be sitting by your fireside.— After leaving the Galapagos, that land of Craters, we enjoyed the prospect, which some people are pleased to term sublime, of the boundless ocean for five & twenty entire days. At Tahiti, we staid 10 days, & admired all the charms of this almost classical Island.— The kind simple manners of the half civilized natives are in harmony with the wild, & beautiful scenery.—

   I made a little excursion of three days into the central mountains. At night we slept under a little house, made by my companions from the leaves of the wild Banana.— The woods cannot of course be compared to the forests of Brazil; but their kindred beauty was sufficient to awaken those most vivid impressions made in the early parts of this voyage.— I would not exchange the memory of the first six months, not for five times the length of anticipated pleasures. . . .

But I must return to Tahiti, which charming as it is, is stupid when I think about all of you.— The Captain & all on board (whose opinions are worth anything) have come to a very decided conclusion on the high merit of the Missionaries.— Ten days no doubt is a short time to observe any fact with accuracy, but I am sure we have seen that much good has been done & scarcely anyone pretends that harm has ever been effected. It was a striking thing to behold my guides in the mountain, before laying themselves down to sleep, fall on their knees & utter with apparent sincerity a prayer in their native tongue. In every respect we were delighted with Tahiti, & add ourselves as one more to the list of the admirers of the Queen of the Islands.—

Again we consumed three long weeks in crossing the Sea to New Zealand, where we shall stay about 10 days.— I am disappointed in New Zealand, both in the country & in its inhabitants. After the Tahitians, the natives, appear savages. The Missionaries have done much in improving their moral character & still more in teaching them the arts of civilization. It is something ⟨to⟩ boast of, that Europæans may here, amongst men who, so lately were the most ferocious savages probably on the face of the earth, walk with as much safety as in England. We are quite indignant with Earle's book,[2] beside extreme injustice it shows ingratitude.— Those very missionaries, who are accused of coldness, I know without doubt that they always treated him with far more civility, than his open licentiousness could have given reason to expect.— I walked to a country mission, 15 miles distant & spent as merry & pleasant an evening with these *austere* men, as ever I did in my life time.[3]

I have written thus much about the Missionaries, as I thought it would be a subject, which would interest you.— I am looking forward with more pleasure to seeing Sydney, than to any other part of the voyage.— our stay there will be very short, only a fortnight; I hope however to be able to take a ride some way into the country.— From Sydney, we proceed to King George's sound & so on as formerly planned. Be sure, not to forget to have a letter at Plymouth on or rather before the 1st. of August. . . .

How glad I shall be, when I can say, like that good old Quarter Master, who entering the Channel, on a gloomy November morning, exclaimed, "Ah here there are none of those d——d blue skys" . . . Give my most affectionate love to my Father, Erasmus Marianne & all of you. Goodbye my dear Caroline | Your's | C. Darwin . . .

To Caroline Darwin 29 April 1836

Port Lewis, Mauritius.
April 29th. 1836.

My dear Caroline,

We arrived here this morning; as a Ship sails for England tomorrow, I will not let escape the opportunity of writing. But as I am both tired & stupid, my letter will be equally dull. I wrote from Sydney & Hobart town, after leaving the latter place, we proceeded to King Georges Sound. I did not feel much affection for any part of Australia; & certainly, nothing could be better adapted, than our last visit, to put the finishing stroke to such feelings.—

We then proceeded to the Keeling Is^{ds}.— These are low lagoon Is^{ds}. about 500 miles from the coast of Sumatra.— I am very glad we called there, as it has been our only opportunity of seeing one of those wonderful productions of the Coral polypi.— The subject of Coral formation has for the last half year, been a point of particular interest to me. I hope to be able to put some of the facts in a more simple & connected point of view, than that in which they have hitherto been considered. The idea of a lagoon Island, 30 miles in diameter being based on a submarine crater of equal dimensions, has alway appeared to me a monstrous hypothesis.[4]

From the Keeling Id we came direct to this place. All which we have yet seen is very pleasing. The scenery cannot boast of the charms of Tahiti & still less of the grand luxuriance of Brazil; but yet it is a complete & very beautiful picture. But, there is no country which has now any attractions for us, without it is seen right astern, & the more distant & indistinct the better. We are all utterly home sick; I feel sure there is a wide difference between leaving one's home to reside for five years in some foreign country, & in wandering for the same time. There is nothing, which I so much long for, as to see any spot & any object, which I have seen before & can say I will see again.— Our heads are giddy, with such a constant whirl. The Capt, continues to push along with a slack rein & an armed heel.— thank Heaven not an hour has lately been lost, or will again be lost.

It is probable, if we escape the heavy gales off the Cape, we may reach England 8 weeks after you receive this letter. Our course beyond the Cape & St Helena is not certain; I think it will end in touching at Bahia on the coast of Brazil. With what different sensations I shall now view that splendid scene, from formerly. Then I thought an hour of such existence would have been cheaply purchased with an year of ordinary life, but now one glimpse of my dear home, would be better than the united kingdoms, of all the glorious Tropics. Whilst we are at sea, & the weather is fine, my time passes smoothly, because I am very busy. My occupation consists in rearranging old geological notes: the rearranging generally consists in totally rewriting them. I am just now beginning to discover the difficulty of expressing one's ideas on paper. As long as it consists solely of description it is pretty easy; but where reasoning comes into play, to make a proper connection, a clearness & a moderate fluency, is to me, as I have said, a difficulty of which I had no idea.—

I am in high spirits about my geology.—& even aspire to the hope that, my observations will be considered of some utility by real geologists. I see very clearly, it will be necessary to live in London for a year, by which time with hard work, the greater part, I trust, of my materials will be exhausted. Will you ask Erasmus to put down my name to the Whyndam or any other club; if, afterwards, it should be advisable not to enter it, there is no harm done. The Captain has a cousin in the Whyndam, whom he thinks, will be able to get me in.— Tell Erasmus to turn in his mind, for some lodgings with good big rooms in some vulgar part of London.— Now that I am planning about England, I really believe, she is not at so hopeless a distance.— Will you tell my Father I have drawn a bill of 30£.— The Captain is daily becoming a happier man, he now looks forward with cheerfulness to the work which is before him. He, like myself, is busy all day in writing, but instead of geology, it is the account of the Voyage. I sometimes fear his "Book" will be rather diffuse, but in most other respects it certainly will be good: his style is very simple & excellent. He has proposed to me, to join him in publishing the account, that is, for him to have the disposal & arranging of my journal & to mingle it with his own. Of course I have said I am perfectly willing, if he wants materials; or thinks the chit-chat details of my journal are any ways worth publishing. He has read over the part, I have on board, & likes it.—

I shall be anxious to hear your opinions, for it is a most dangerous task, in these days, to publish accounts of parts of the world, which have so frequently been visited. It is a rare piece of good fortune for me, that of the many errant (in ships) Naturalists, there have been few or rather no geologists. I shall enter the field unopposed.— I assure you I look forward with no little anxiety to the time when Henslow, putting on a grave face, shall decide on the merits of my notes. If he shakes his head in a disapproving manner: I shall then know that I had better at once give up science, for science will have given up me.— For I have worked with every grain of energy I possess.— But what a horridly egotistical letter, I am writing; I am so tired, that nothing short of the pleasant stimulus of vanity & writing about one's own dear self would have sufficed.— I have the excuse, if I write about my self, Heaven knows I think enough about all of you. . . .

give my most affectionate love to my Father & all | My dearest Caroline | Your affectionate brother | Chas. Darwin.

To Susan Darwin 4 August [1836]

Bahia, Brazil
August 4th.

My dear Susan

I will just write a few lines to explain the cause of this letter being dated on the coast of S. America.— Some singular disagreements in the Longitudes, made Capt. F. R. anxious to complete the circle in the Southern hemisphere, & then retrace our steps by our first line to England.— This zig-zag manner of proceeding is very grievous; it has put the finishing stroke to my feelings. I loathe, I abhor the sea, & all ships which sail on it. But I yet believe we shall reach England in the latter half of October. . . .

Both your letters were full of good news:— Especially the expressions, which you tell me Prof: Sedgwick used about my collections.—⁵ I confess they are deeply gratifying.— I trust one part at least will turn out true, & that I shall act, as I now think.—that a man who dares to waste one hour of time, has not discovered the value of life.— Prof. Sedgwick men⟨tionin⟩g my name at all gives me hopes that he will assist me with his advice; of which in many geological questions, I stand much in need. . . .

Farewell, my very dear Susan & all of you.. Goodbye | C. Darwin—

To Josiah Wedgwood II [5 October 1836]

[Shrewsbury]

My dear Uncle

The Beagle arrived at Falmouth on Sunday evening, & I reached home late last night. My head is quite confused with so much delight, but I cannot allow my sisters to tell you first, how happy I am to see all my dear friends again. I am obliged to return in three or four days to London, where the Beagle will be paid off, & then I shall pay Shrewsbury a longer visit. I am most anxious once again to see Maer, & all its inhabitants, so that in the course of two or three weeks, I hope in person to thank you, as being my first Lord of the Admiralty. I am so very happy I hardly know what I am writing.

Believe me, Your most affectionate nephew | Chas. Darwin . . .

1837

[Soon after his return home, Darwin made several trips to London. His letters to Henslow and the specimens he had sent from the *Beagle* had already established him in scientific circles. He dined with Charles Lyell and Richard Owen, was elected to the Geological Society, and arranged to have his collections examined by specialists. In December he took lodgings in Cambridge to arrange the specimens Henslow had stored for him.]

To Caroline Darwin   27 February 1837

[Cambridge]
Monday, Feby 27th. 1837.

My dear Caroline

It is nearly twelve o'clock, but before going to bed I will write my last letter from Cambridge.— I have just been reading a short paper to the Philosoph. Soc^{y}. of this place, and exhibiting some specimens & giving a verbal account of them. It went off very prosperously & we had a good discussion in which Whewell & Sedgwick took an active part.— Sedgwick has just come from Norwich & we have been drinking tea with him.— He always enquires very particularly about my Father and all of you.— I really sometimes think he will go mad; he is so very absent & odd, but a more high-minded man does not anywhere exist. On Friday morning I migrate. My Cambridge life is ending most pleasantly.— You enquired in y^{r}. last letter about Lyell's Speech; very little was said about me, as of course he could only allude to published accounts.— But if you think it worth while I will send it down to you,—(and at the same time the Missionary paper, which has arrived from C. of Good Hope.)[1] I heard from Lyell yesterday, he says it will be published in two or three days.— He wants me to be up on Saturday for a party at M^{r}. Babbage, who has sent me a card for his parties this season   Lyell says Babbage's parties are the best in the way of literary people in London—and that there is a good mixture of pretty women—

You tell me you do not see what is new in Sir J. Herschell's idea about the chronology of the old Testament being wrong.— I have used the word Chronology in dubious manner, it is not to the days of Creation which he refers, but to the lapse of years since the first man made his wonderful appearance on this world— As far as I know everyone has yet thought that the six thousand odd years has been the right period but Sir J. thinks that a far greater number must have passed since the Chinese, the [*space left in copy*],[2] the Caucasian languages separated from one stock. . . .

Yours affectionly, | C. Darwin.

To W. D. Fox   [12 March 1837]

[43 Great Marlborough Street]

Sunday Evening

My dear Fox

It is a long time since I wrote to you, from Cambridge, but I was determined to wait till I was fairly settled, which however I can hardly say I am yet, but on Tuesday I go into lodgings, at No^r 36 Gr^t. Marlborought St. which I have taken for the year.— I am at present in my brothers house no 43.— It is very pleasant our being so near neighbours.— My residence at Cambridge was rather longer, than I expected, owing to a job, which I determined to finish there, namely looking over all my geological specimens was finished.— Cambridge yet continues a very pleasant, but not so half so merry a place as before.— To walk through the courts of Christ Coll: and not know an inhabitant of a single room gave one a feeling half melancholy.— The only evil I found in Cambridge, was its being too pleasant; there was some agreeable party or another every evening, and one cannot say one is engaged with so much impunity there as in this great city.— It is a sorrowful, but I fear too certain truth, that no place is at all equal, for aiding one in Natural History pursuits, to this odious dirty smokey town, where one can never get a glimpse, at all, that is best worth seeing in nature.—

In your last letter you urge me, to get ready *the* book. I am now hard at work and give up every thing else for it. Our plan is as follows.— Capt. FitzRoy writes two volumes, out of the materials collected during both the last voyage under Capt. King to T. del Fuego and during our circumnavigation.— I am to have the third volume, in which I intend giving a kind of journal of a naturalist, not following however always the order of time, but rather the order of position.— The habits of animals will occupy a large portion, sketches of the geology, the appearance of the country, and personal details will make the hodge-podge complete.— Afterwards I shall write an account of the geology in detail, and draw up some Zoological papers.— So that I have plenty of work, for the next year or two, and till that is finished I will have no holidays.— Do you recollect telling me, the new ostrich should be called "*darwinii*". By an odd chance Mr Gould has actually so named it!— We are going to read a paper to the Zoological on Tuesday about it. . . .

Good Bye, Dear Fox. | Yours affectionly, C Darwin

To Caroline Darwin [19 May – 16 June 1837]
36. Great Marlborough St.
My dearest Caroline,

. . . Will you ask my Father whether he did not say that floods of the Severn from Snow were less muddy & less destructive to the soil than those from heavy rains, or was it vice versa or what was it? Is it in the Zoonomia or notes to Botanic Garden where there is a passage about acquired instincts, such as crows learning Guns are dangerous—[3] I have been much interested at finding so many cases where ships with all their crew in good health have yet caused strange contagious disorders at Station Islands in the Pacific— M^r. Williams, a missionary, boldly asserts that the first mingling, where both are healthy of the European race with the natives of distant climes always produces disease. I am very much inclined to suspect that there is some such mysterious law connected with the destruction of the Aborigines in both Americas—Cape of Good Hope—Australia & Polynesia.— I recollect years ago my Father mentioning (from Macculloch) some little Island where the people had influenza whenever a ship arrived—& such was thought to be explained by Vessels always arriving with certain wind. Will you ask my Father what my memory is alluding to.— Again, are there not cases where people packed together have produced most deleterious contagion *without themselves being affected!.* — Was not the "black Assize of Oxford" such a case? What year was it? Have I not heard my Father mention some other parallel instance.— Now will you be a good lady & look at Ellis' Polynesian Researches,[4] & see if he does not at Tahiti make some remarks about the belief that Capt. Cook's visits produced some kinds of illness— There is some ridiculous story about humpbacks, but perhaps some truth in other cases after all.— Ellis is so well divided into Chapters, that you will easily find a probable chapter to read, if not don't trouble yourself— Will you ask my Father whether he has ever heard of any experiment to try, as the puncture from instruments in dissecting a man is so fatal to another man whether a dead dog for instance, would so act to living dog.— Was there ever such an odd string of questions? The Governor will think I am gone mad;[5] I cannot in letter show connection of questions . . .

Give my most affectionate love to my Father & say I hope the questions wont plague him. Good bye. C. Darwin . . .

To Francis Beaufort 16 June 1837

June 16th. 1837

My dear Sir

Having accompanied, as you are aware, Captain FitzRoy round the world, I have been enabled during the five years to make extensive collections in the various departments of Natural History. Several scientific gentlemen having examined them, are of opinion that the results should be published; but finding myself unable to support the expense of the numerous engravings, which would be necessary, I have ventured to hope for the assistance of Government, in a similar manner to that afforded to Sir J. Franklin, D^r. Richardson and to others, so as to be enabled to produce a work, which would be creditable to the country. I am emboldened to trust that this request will not at least be thought presumptious, more especially, when I state that the entire expense, even to the purchase of materials for the preservation of the specimens, together with a salary for an assistant has been willingly defrayed by myself. I may add that the whole collection has already been, or will hereafter be distributed to the public museums, where they will be of acknowledged service. I have subjoined the opinion of the Presidents of the three Learned Societies, respecting the utility of this publication, if illustrated by engravings, and any assistance, which Government might afford, should exclusively be employed on the new and undescribed species. Other naturalists and myself are of opinion that 150 plates are necessary, and that the expense of drawing and engraving would amount to about a thousand pounds, but required only by instalments during a year and a half.

Pray tell me to what quarter I should apply, and | Believe me, dear Sir | Yours most truly obliged | Chas. Darwin . . .

[Enclosure]

Being fully impressed with the importance and novelty of the collections in Natural History brought home by M^r. Darwin in H.M.S

Beagle, we are of opinion that their publication on some uniform plan, illustrated by engravings, would be highly advantageous to Natural Science.

Somerset | Derby | W. Whewell

To W. D. Fox   7 July [1837]

36 Grt. Marlborough St.—
Friday July 7th.

My dear Fox

It is a very long time since I have heard any news of you; why have you not written to tell me how you are going on?— Are you turned idle, or do you think I am too full of South American bird beasts and fishes to care about old friends.— I returned last night from a flying visit of a eight days to Shrewsbury; & I have now got a piece of news to tell you, which I am sure you will be interested about. Caroline is going to be married to Jos Wedgwood. ... I do not know whether you recollect him, he is the eldest son.— He is a very quiet grave man, with very much to respect & like in him, but I wish he would put himself forward more. He has a most wonderful deal of information, & is a very superior person; but he has not made the most of himself.— I am very glad of the marriage for Caroline's sake, as I think she will be a very happy person, especially if she has children, for I never saw a human being so fond of little crying wretches, as she is. But I am an ungrateful dog to speak this way, for she was a mother to me, during all the early part of my life.— And I forget I must not talk to *you* of;—crying little wretches— You will not guess that I mean such little angels as all children doubtless are. ...

I gave myself a holiday, and a visit to Shrewsbury, as I finished my journal, I shall now be very busy in filling up gaps & getting it quite ready for the press, by the first August.— I shall always feel respect for every one who has written a book, let it be what it may, for I had no idea of the trouble, which trying to write common English could cost one.— And alas there yet remains the worst part of all correcting the press.— as soon as ever that is done I must put my shoulder to the wheel & commence at the geology.— I have read some short papers to the geological Soc, & they were favourably received by the great guns, & this gives me much confidence, & I hope not a very great deal of vanity; though I confess I feel too often like a peacock admiring his tail.— I never expected that my geology would ever have been worth the consideration of such men, as Lyell, who has been to me, since my return a most active friend. ...

I often think of our old Entomological walks,— at this moment I can see a part of a wood (famous for big fungi & little jumping beetles (anaspis? orchesia)) near Osmaston, as plain as if we had been sweeping there a month ago instead of some seven long years.— And many a good day in Cambridge has left, & will I think ever leave, as bright a picture in my mind, as any I am capable of enjoying. . . . God Bless you dear Fox.— Ever yours affectionaly | C. Darwin.

[*'In July opened first note Book on 'transmutation of Species'.*— *Had been greatly struck from about month of previous March—on character of South American fossils—& species on Galapagos Archipelago.*— *These facts origin (especially latter) of all my views.'* (*Journal*, 1837.)]

To Charles Lyell   30 July 1837

36 G.^t Marlboro' St^t.

July 30th. 1837.—

I believe there are 27 land birds from the Galapagos, all new except one, (a species of very wide range) yet all of an American form, some north, some south, Now as the Galapos is on the Equator is not this curious— Reptiles the same— Has your late work at shells startled you about the existence of species? I have been attending a **very** little to species of birds, & the passages of forms, do appear frightful— every thing is arbitrary; no two naturalists agree on any fundamental idea that I can see. I had a most interesting morning with Owen (who is gone to rest for a month in the N. of England) at the Coll. of Surgeons— We made out the rem.^s of 11 or 12 great animals, besides these some rodents, one of wh. is a distinct species, but most strictly S. American genus. At Bahia Blanca there were no less than *five great* Edentatas! what could these monsters have fed upon— I am well convinced like the present Armadillos they lived on land nearly desert— I have worked out the *non* relation of bulk of animals & luxuriance of vegetation, & I have been perfectly astonished at some of the facts given me by Dr Smith.[6] If it would be any satisfaction to you I think it could be proved rhinoceroses live upon air, certain it is they must be light feeders. What will you say to the tusk of a boar like the African species being imbedded with the Edentata. Lastly I am sure when you read my evidence (& see the tooth) you will be as much convinced as I am that a horse was formerly common on the Pampas as at the present day. What an extraordin.^y mystery it is, the cause of the death of these numerous animals, so recently, & with so little physical change.—

[Incomplete]

[In August Darwin was granted £1000 towards publication of the *Zoology of the Beagle* by the chancellor of the Exchequer. To describe the specimens, he enlisted Richard Owen for *Fossil Mammalia*, George Robert Waterhouse for *Mammalia*, John Gould for *Birds*, Leonard Jenyns for *Fish*, and Thomas Bell for *Reptiles*. Darwin superintended the work, supplying from his field notes descriptions of geographical and geological data, of habitats and habits.]

To J. S. Henslow   14 October [1837]

[Shrewsbury]
October 14th.—

My dear Henslow

... I am much obliged to you for your message about the Secretaryship:[7] I am exceedingly anxious for you to hear my side of the question, & will you be so kind as afterwards to give me your fair judgment.— The subject has haunted me all summer. I am unwilling to undertake the office for the following reasons.— 1st. My entire ignorance of English geology, a knowledge of which would be almost necessary in order to shorten many of the papers before reading them, before the Society, or rather to know what parts to skip— Again my ignorance of all languages; & not knowing how to pronounce even a *single* word of French,—a language so perpetually quoted. It would be disgraceful to the Society to have a Secretary who could not read French. 2^d. The loss of time. Pray consider, that I shall have to look after the artists, superintend & furnish materials for the government work, which will come out in parts, & which must appear regularly. All my geological notes are in a very rough state, none of my fossil shells worked up, and I have much to read. I have had hopes by giving up society & not wasting an hour, that I should be able to finish my geology in a year and a half, by which time the description of the higher animals by others would be completed & my whole time would then necessarily be required to complete myself the description of the invertebrate ones. If this plan fails, as the government work must go on, the geology would necessarily be deferred till probably at least three years from this time.

In the present state of the science a great part of the utility of the little I have done, would be lost, and all freshness and pleasure quite taken from me. I know from experience the time required to make abstracts, *even* of my own papers, for the Proceedings. If I was secretary & had to make double abstracts of each paper, studying them before reading, and attendance would *at least* cost me three days (& often more) in the fortnight. There are likewise other accidental and contingent losses of time.— I know Dr Royle found the office consumed much of his time.— If by merely giving up any amusement, or by working harder than I have done, I could save time, I would undertake the secretaryship, but I appeal to you, whether, with my slow manner of writing,—with two works in hand,—and with the certainty, if I cannot complete the geological part within a fixed period, that its publication must be retarded for a very long time, whether any Society whatever has any claim on me for three day's disagreeable work every fortnight. I cannot agree that it a *duty* on my part, as a follower of science, as long as I devote myself to the completion of the work I have in hand, to delay that by undertaking what may be done by any person, who happens to have more spare time than I have at present. Moreover so **early** in my scientific life, with so very much as I have to learn, the office, though no doubt a great honour &c for me, would be the more burdensome. Mr Whewell, I know very well, judging from himself, will think I exaggerate the time the Secretaryship would require, but I absolutely know, the time, which with me the simplest writing consumes. I do not at all like appearing so selfish as to refuse Mr Whewell, more especially as he has always shewn, in the kindest manner, an interest in my affairs.— But I cannot look forward with even tolerable comfort to undertaking an office, without entering on it heart and soul, and that would be impossible with the Government work and the geology in hand.

My last objection, is that I doubt how far my health will stand, the confinement of what I have to do without any additional work. I merely repeat, that you may know I am not speaking idly, that when I consulted Dr Clark in town, he at first urged me to give up entirely all writing and even correcting press for some ⟨w⟩eeks. Of late, anything which flurries me completely knocks me up afterwards and brings on a b⟨ad⟩ palpitation of the heart. Now the Secretaryship would be a periodical source of more annoying trouble to me, than all the rest of the fortnight put together. In fact till I return to town and see how I get on, if I wished the office ever so much, I *could* not say I would positively undertake it.

I beg of you to excuse this very long prose all about myself; but the point is one of great interest.— I can neither bear to think myself very selfish and sulky, nor can I see the possibility of my taking the secretaryship with making a sacrifice of all my plans, and a good deal of comfort.— If you see Whewell would you tell him the substance of this letter; or if he will take the trouble, he may read it.— My dear Henslow, I appeal to you in loco parentis,—pray tell me what you think. But do not judge me by the activity of mind, which you and a few others possess, for in that case, the more different things in hand, the pleasanter the work but, though I hope I never shall be idle, such is not the case with me.

Ever, dear H. | Y^T. most truly, | C. Darwin

[CD eventually yielded to William Whewell's insistence. He was elected one of the two secretaries of the Geological Society on 16 February 1838 and served until 19 February 1841.]

1838

To Susan Darwin   [1 April 1838]

                                                    Sunday Evening

My dear Granny

... I went to the Captains yesterday evening to drink tea.— it did one good to hear Mrs FitzRoy talk about her baby; it was so beautiful & its little voice was such charming music.— The Captain is going on very well,—that is for a man, who has the most consummate skill in looking at everything & every body in a perverted manner.— He is working very hard at his book which I suppose will really be out in June.— I looked over a few pages of Captain Kings Journal: I was absolutely forced against all love of truth to tell the Captain that I supposed it was very good, but in honest reality, no pudding for little shool boys, ever was so heavy.— It abounds with Natural History of a very trashy nature.— I trust the Captain's own volume will be better.—

I have been riding very regularly for the last fortnight, & it has done me a wonderful deal of good.— I have not been so thoroughily well, since eating two dinner a day at Shrewsbury, & increasing in weight in due proportion.—

Two days since, when it was very warm, I rode to the Zoological Society, & by the greatest piece of good fortune it was the first time this year, that the Rhinoceros was turned out.— Such a sight has seldom been seen, as to behold the rhinoceros kicking & rearing, (though neither end reached any great height) out of joy.— it galloped up & down its court surprisingly quickly, like a huge cow, & it was marvellous how suddenly it could stop & turn round at the end of each gallop.— The elephant was in the adjoining yard & was greatly amazed at seeing the rhinoceros so frisky: He came close to the palings & after looking very intently, set off trotting himself, with his tail sticking out at one end & his trunk at the other,—squeeling & braying like half a dozen broken trumpets.— I saw also the Ourang-outang in great perfection: the keeper showed her an apple, but would not give it her, whereupon she threw herself on her back, kicked & cried, precisely like a naughty child.— She then looked very sulky & after two or three fits of pashion, the keeper said, "Jenny if you will stop bawling & be a good girl, I will give you the apple.— She certainly understood every word of this, &, though like a child, she had great work to stop whining, she at last succeeded, & then got the apple, with which she jumped into an arm chair & began eating it, with the most contented countenance imaginable.—

So much for Monkey, & now for Miss Martineau, who has been as frisky lately ⟨as⟩ the Rhinoceros.— Erasmus has been with her noon, morning, and night:— if her character was not as secure, as a mountain in the polar regions she certainly would loose it.— Lyell called there the other day & there was a beautiful rose on the table, & she coolly showed it to him & said "Erasmus Darwin" gave me that.— How fortunate it is, she is so very plain; otherwise I should be frightened: She is a wonderful woman: when Lyell called, he found Rogers, L^d. Jeffrys, & Empson calling on her.— what a person she is thus to collect together all the geniuses.— Old Rogers seems to a warm admirer of hers.— He says her laugh is so charming, it is "like tickling a child in a cradle." Was there ever such a simile.—a pretty little baby indeed. ...

Love to my father | C Darwin.

[During the next months, besides working on the *Zoology*, CD began to write up his geological field notes and continued to gather data for his notebooks on the transmutation of species.

On 9 November, CD set out for Maer, the Wedgwood residence in Staffordshire, with the intention of proposing to his cousin, Emma Wedgwood. On Sunday the 11th, '*The day of days!*', he was accepted (*Journal*, 1838).]

To Emma Wedgwood   [14 November 1838]

Shrewsbury
Wednesday Morning

My dear Emma

Marianne & Susan will have told you what joy and happiness the news gave all here. We have had innumerable cogitations and *geese*[1] after the cogitations; and the one conclusion, I exult in, is that there never was anybody so lucky as I have been, or so good as you.— Indeed I can assure you, many times, since leaving Maer, I have thought how little I expressed, how much I owe to you; and as often as I think this, I vow to try to make myself good enough somewhat to deserve you.—

I hope you have taken deep thought about the sundry knotty points you will have to decide on.— We must have a great deal of talk together, when I come back on Saturday.— do have a fire in the Library—it is a such a good place to have some quiet talks together.— The question of houses,—suburbs versus central London,—rages violently around each fire place in this house.— Suburbs have rather the advantage at present; & this, of course, rather inclines one to seek out the argument on the other side.— The Governor gives much good advice to live, wherever it may be, the first year prudently & quietly. My chief fear is, that you will find after living all your life with such large & agreeable parties, as Maer only can boast of, our quiet evenings dull.— You must bear in mind, as some young lady said, 'all men are brutes', and that I take the line of being a solitary brute, so you must listen with much suspicion to all arguments in favour of retired places. I am so selfish, that I feel to have you to myself, is having you so much more completely, that I am not to be trusted. Like a child that has something it loves beyond measure, I long to dwell on the words *my own* dear Emma.— as I am writing, just as things come uppermost in my mind, I beg of you not to read my letters to anyone, for then I can fancy, I am sitting by the side of my own dear future wife, & to her own self, I do not care what nonsense I talk:—so let me have my way, & scribble, without caring whether it be sense or nonsense.—

I had a letter from Caroline yesterday, full of kindest & tenderest expressions towards us both.— I do not mean to tell anyone in Shropshire, till I leave; but I have written to Erasmus, & I am well sure he will most heartily congratulate. My father echos & reechos Uncle Jos' words 'you have drawn a prize!' Certainly no man could by possibility receive a more cordial welcome, that I did from everyone at Maer on Monday morning.— My life has been very happy & very fortunate and many of my pleasantest remembrances are mingled up with scenes at Maer, & now it is crowned.— My own dear Emma, I kiss the hands with all humbleness and gratitude, which have so filled up for me the cup of happiness— it is my most earnest wish, I may make myself worthy of you.

Good bye | Most affectionately Yours | Chas Darwin

I would tear this letter up & write it again, for it is a very silly one, but I cant write a better one.—

Since writing the former part, the Post has brought in your own dear note to Katty. You tell me to be a good boy, & so I must be,—but let me earnestly beg of you not to make up your mind, in a hurry.— You say **truly** Elizabeth never thinks of herself, but there is another person, who never thinks of herself, but now she has to think of two people,—& I am, thank Heaven for it, that other person.— You must be absolute arbitress, but do dear Emma, remember life is short, & two months is the sixth part of the year, & that year, the first, from which for my part, things shall hereafter date. Whatever you do will be right,—but it will be *too* good to be unselfish for me, until I am part of you, dearest Emma.

good bye

To Emma Wedgwood   [27 November 1838]

Athenæum
Tuesday night

My dear Emma

I have taken a *large sheet* of paper, … & mean to have a most comfortable prose with you. …

I positively can do nothing, & have done nothing this whole week; but think of you & our future life.— you may then, well imagine how I enjoy seeing your handwriting. I should have written yesterday but waited for your letter: pray do not talk of my waiting till I have time for writing or inclination to do so.— it is a very high enjoyment to me, as I cannot talk to you, & feel your presence, by having your own dear hand within mine.—

I will now narrate my annals: on Saturday I dined with the Lyells, & spent one of the pleasantest evenings I ever did in my life. Lyell grew quite audacious, at the thoughts of having a married geological companion, & proposed going to dine at the Athenæum together & leaving our wives at home.— Poor man, he would as soon '*eat his head*,' as do such an action, whilst I feel as yet as bold as a lion. We had much geological & economical talk,—the latter very profitable.— By the way if you will take my advice, you will not think of reading the Elements, for depend upon it you will hereafter have plenty of geology: act on the same principle, which makes me take as much snuff as possible, before the 24th.—to make an unfeeling joke, let us both be happy as long as we can.

On Sunday evening Erasmus took me to drink tea with the Carlyles; it was my first visit.— One must always like Thomas, & I felt particularly well towards him, as Erasmus had told me he had pro-pounded that a certain lady, was one of the nicest girls he had ever seen.— Jenny sent some civil messages to you, but which from the effects of an hysterical sort of giggle were not very intelligible. It is high treason, but I cannot think that Jenny is either quite natural or lady-like. . . .

And now for the great question of houses. Erasmus & myself have taken several very long walks; & the difficulties are really frightful. Houses are very scarce & the landlords are all gone mad. they ask such prices.— Erasmus takes it to heart, even more than I do, & declares I ought to end all my letters to you "Your's inconsolably."— This day I have given up to deep cogitations regarding the future, in as far as houses are concerned. It would take up too much paper to give all the pro's & cons; but I feel sure, that a central house would be best for both of us, for two or three years.— I am tied to London, for rather more than that period; & whilst this is the case, I do not doubt it is wisest to reap *all* the advantages of London life: more especially as every reason will urge us to pay frequent visits to *real* country, which the suburbs never afford. After the two or three years are out, we then might decide whether to go on living in the same house or suburbs, supposing I should be tied for a little longer to London & ultimately to decide, whether the pleasures of retirement & country, (gardens, walks, &c) are preferable to society &c &c. It is no use thinking of this question at present & I repeat, I do not doubt, your first decision was right: let us make the most of London, whilst we are compelled to be there: the case would be different, if we were deciding for life, for then we might wish to possess the advantages both of country & town, though both in a lesser degree, in the suburbs.— Tell me what you think of this reasoning.— I am glad to hear you are oscillating in opinion, as you will make all the better judge.— With respect to what *part* of central London, I clearly see, the possibility of obtaining a house *must* settle the question.— After many long walks, Erasmus & myself are driven to the conviction, that our only resource will be in the streets or squares, near Russell Square.—

After much deliberate talk, (especially with the Lyells), I have no doubt, that our best plan will be to furnish slowly a house for ourselves.— it will be far more economical both in money & time; but not in *comfort* just at first.— Will you rough it a little at first? Again I clearly see we shall be obliged to give at least 120£ for our

house; if not a little more: The most promising one Erasmus have yet
seen is in Torrington Square for 120£: the one in Tavistock square
was 150£; one in Bedford place 145£.— I will steadily go on looking
& pondering: I believe I have good reason for the points, that I have
spoken on; but I wish much to hear all suggestions from you, and
mind be not, like yourself & Elizabeth too unselfish.— Regents Park
is a failure. I have been all round it.

Until yesterday I intended to have paid Maer a visit on Thursday
week—the day after the geolog Soc. but yesterday I heard of the
death of the mother of Mʳ Owen, who was to write the next number
of the Government Work, which now he will not probably be able
to do, & I am put to my wit's ends to get some other number ready.
How long this will delay me, I can hardly yet tell. I hope most
earnestly not long, for I am impatient to see you again. It is most
provoking, I can not settle down to work in earnest, just at the very
time I most want to do so.— There is the Appendix of the Journal &
half a dozen things, besides this unlucky number all waiting my good
pleasure.— every night I make vows & break them in the morning.
I do long to be seated beside of you, again, in the Library; one can
then almost feel anticipation, the happiness to come.

I have just read your letter over again for the fifth time. my own
dear Emma, I feel as if I had been guilty of some very selfish action
in obtaining such a good dear wife—with no sacrifice at all on my
part—as I have said before I must try & make a very dutiful &
grateful husband.

Believe me, dear Emma. Most affectionately yours. C. Darwin ...

[Emma came to London for a fortnight to help with house hunting and other
preparations for the wedding. A house was found and rented at 12 Upper Gower
Street by the end of December. They were married at Maer on 29 January 1839.
The couple left immediately for London, where CD worked on *Coral reefs*.]

1839–1843

To J. S. Henslow [10 November 1839]

12 Upper Gower St
Sunday Evening

My dear Henslow

... I believe you have received a message I sent you saying that Humboldt in a letter to Me expresses at great length his **vivid** *regret* that M. Henslow has not been able to describe the species, or even characterize the genera of the very curious collection of plants from Galapagos.— Do think once again of making one paper on the Flora of these islands—like Roxburgh on St. Helena, or Endlicher on Norfolk Isl^d.— I do not think there will often occur opportunities of drawing up a monograph of more interest.— if your descriptions are frittered in different journals, the general character of the Flora never will be known, & foreigners, at least, will not be able to refer to this & that journal for the different species— But you are the best judge.—

I have been lately reading some remarks on the geograph. distrib. of plants & I am very curious to have a paper at some time from you on the general character of the Flora of T. del. Fuego & especially of the Alpine Flora⟨.⟩ The one point of land, which projects so far into temper⟨ate⟩ countries ought to be characterized by very peculiar forms in relation to the northern hemisphere.— Robert Brown has a very large & I believe perfect collection from Tierra del Fuego, which I daresay he would allow you to undertake, if you chose, as it has been in his possession about nine years.— I do not believe, anyone has published any general account of the Flora of Patagonia,— small as my collection is,—it gives I am sure a very fair notion of the Flora—& the climate being so peculiar, I cannot fancy anything more remarkable than the contrasts of its Flora with that of Tierra del Fuego, countries so near to each other.— Do think of these points.

... Dont forget to bear in mind, as you said you would for me, to notice any facts either hostile or corroboratory of my notion of all plants occasionally impregnating each other.— I keep on steadily collecting every sort of fact, which may throw light on the origin & variation of species.—

My wife sends her kind regards to you— we shall always be **most happy** to see you here, & the oftener the better & we hope some time to accept your kind invitation for Hitcham.

Good Bye | My dear Henslow | Ever yours | Chas. Darwin

To Emma Darwin   [5 April 1840]

Shrewsbury
Sunday.

My dear Emma.—

You are a good old soul for having written to me so soon.— I, like another good old soul, will give you an account of my proceedings from the beginning.— At the station I met Sir F. Knowles, but was fortunate enough to get in a separate carrigae from that chatterbox. In my carriage, there was rather an elegant female, like a thin Lady Alderson, but so virtuous that I did not venture to open my mouth to her. She came with some female friend, also a lady & talked at the door of the carriage in so loud a voice, that we all listened with silent admiration. It was chiefly about family prayers, & how she always had them at $\frac{1}{2}$ past ten not to keep the servants up. She then charged her friend to write to her either on Saturday night or Monday morning, Sunday being omitted in the most marked manner.— Our companion answered in the most pious tone, "Yes Eliza I will write either on Saturday night or on Monday morning."— as soon as we started our virtuous female pulled out of her pocket a religious tract in a black cover, & a very thick pencil,— she then took off her gloves & commenced reading with great earnestness & marking the best passages with the aforesaid thick lead-pencil.— Her next neighbour was an old gentleman with a portentously purple nose, who was studying a number of the Christian Herald, & his next neighbour was the primmest she Quaker I have often seen.— Was not I in good company?— I never opened my mouth & therefore enjoyed my journey.

At Bermingham, I was kept standing in the office $\frac{3}{4}$ of an hour in doubt, whether I could have a place, & I was so tired, that I regretted much that I took one,— however to my surprise the Journey rested me, & I arrived very brisk at Shrewsbury. In the office at Bermingham, I was aghast to see M^r. J. Hunt, an indomitable proser, taking his place.— He did not know me, as I found by his addressing a chance remark to me, I instantly resolved on the desperate attempt of travelling the whole way incognito.— My hopes were soon cut off by the appearance of M^{rs}. Hunt, whom I shook hands with vast surprise & interest, & opened my eyes with astonishment at M^{r}. Hunt, as if he had dropped from the skies.— Our fourth in the Coach was Mr Parr of Lyth,—an old miserly squire. M^{r}. Hunt opened his battery of conversation,— I stood fire well at first & then pretended to become very sleepy,— the proser became really so, so we had the most tranquil Journey.— Old Parr, the miser, was sadly misused at the Lion, for he had ordered a Fly to take him home, & there was only one, & Mark persuaded the man to take me up first, & gave a hint to the Porters to take a wonderful time in getting old Parr's things off the Coach, so that the poor old gentleman must have thought the Porters & Fly men all gone mad together, so slowly no doubt they did everything, whilst I was driving up with the most surprising alacrity.—

My Father is appearing very well.— I have begun to extract wisdom from him, which I will not now write.—

He does not seem able to form any opinion about your case.—but strongly urges your going on suckling a little for some time, even at the expense of slight headachs.—[1] He says you probably will be able to guess with better chance of truth later about your condition—but that it will be only a guess.— You will be pleased to hear, that he objects to the Baby having medicine for every trifle— He says, *as long as the Baby keeps well*, & the motions appear pretty healthy, he thinks it of little consequence whether it has a dozen or one or even less than one in 24 hours, although he says it is desirable it should be more than once.— He is very strong against gruel, but not against other food.— He thinks there is no occasion to go on with Asses' milk.— But I will tell you all this when I come back.—

I enjoy my visit & have been surprisingly well & have not been sick once.— My Father says I may often take Calomel.— He has recommended me nothing in particular.— I find I am a good deal thinner than I was, weighing less than Erasmus now.— I suspect the Journey & change will do me good— I could not, however, sleep but very little the first night, & I verily believe it was from the lonesomeness of the big bed,—in which respect I have shown much more sentimentality than, it appears you did. I have begun, like a true old Arthur Gride[2] making a small collection & have picked up several nice little things, & have got some receipt for puddings &c

& laid some strong effectual hints about jams, & now you may send
the empty jars when-ever you please. . . .

Good Bye my dear old Titty. I am often thinking about you.
Give my best love to dear old Katty, & believe your affectionate old
Husband | C. D

To W. D. Fox   [25 January 1841]

12 Upper Gower St
Monday

My dear Fox

It is a long time since we have had any communication,— I daresay you will be glad to hear how I am going on, & I wish to hear of
M^{rs}. Fox and yourself.— My strength is gradually, with a good many
oscillations, increasing; so that I have been able to work an hour or
two several days in the week— I have at last to my great joy sent
the last page of M.S. of the Bird Part to the printer.— I am forced
to live, however, very quietly and am able to see scarcely anybody
& cannot even talk long with my nearest relations. I was at one time
in despair & expected to pass my whole life as a miserable useless
valetudinarian but I have now better hopes of myself— You see I
treat you, like a very old good friend, as you have always been to
me, & write at great length about my own poor carcase.

As for news of any kind, I am not in the way to give any to
any body— We have the pleasure, at present, of a visit from Susan,
who is in a very flourishing state of health, which for a Darwin is
something wonderful.—but I will say nothing more about health, &
as a consequence I must say nothing more about any of my Family—
I will just add that Emma expects to be confined in March[3] —a
period I most devoutly wish over— Our little boy is a noble fat little
fellow & my Father has christened him Sir Tunberry Clumsy.—[4]

Pray let me hear soon how you are all going on.— I hope this
very severe winter has not much affected M^{rs}. Fox and that your
lungs have stood it pretty well.— My dear old friend you have much
to support.

Yours affectionly | Ch. Darwin

PS. | If you attend at all to Nat. Hist—I send you this P.S. as a me-
mento, that I continue to collect all kinds of facts, about "Varieties
& Species" for my some-day work to be so entitled—the smallest
contributions, thankfully accepted—descriptions of offspring of all
crosses between all domestic birds & animals dogs, cats &c &c very

valuable— Dont forget, if your half-bred African Cat should die, that I should be very much obliged, for its carcase sent up in little hamper for skeleton.— it or any cross-bred pidgeon, fowl, duck, &c &c will be more acceptable than the finest haunch of Venison or the finest turtle.— Perhaps all this will only bothers you— So I will add no more, except, should you ever have opportunity when in Derbyshire, do enquire for me, from some person you told me of whether offspring of male muscovy & female common duck, resembles offspring of female muscovy & male common— How many hybrid eggs are produced. . . .

To Henry Thomas De la Beche   7 February 1842

(1). What colour are the horses, which have been bred for some generations (without a cross with any foreign blood) in Jamaica, referring, especially to such horses, as have been little taken care of & have run loose.— Have you ever heard of horses of certain colours, having been introduced, whose descendants are now of a different colour.—

(2) The same question with respect to cattle. Are the cattle in half or quite wild herds of the same colour, and marked about the face in the same manner, one with another: that is do they resemble each other closely (like the individuals of those animals which at no time have been domesticated by man) in those points, namely length & curvature of horns size of dewlap, length & fineness of hair, size form & proportions of body, head, & limbs—points, in which different breeds of cattle usually differ from each other.— Please to give a brief description of the half wild cattle.—

(3.) Can you give me any information on the above points (with brief description) regarding dogs, cats, poultry, pigs, goats, run wild in the woods.— If there are any dogs, cats or pigs wild, & if very young ones be caught & reared in a house, do they become quite as tame; and with the same disposition, as the ordinary tame breeds.— If they differ in disposition, & if crossed with ordinary tame breeds, do their offspring still retain any traces of the peculiarities of their half wild parent?—

(4) In your own experience, have you observed any evident deterioration in the character of the wool of imported sheep, after having been bred for a few generations in Jamaica.—
    C. Darwin . . .

[In the summer of 1842, the Darwins decided to leave London to reside in the country. Down House, in Downe, Kent, about sixteen miles south-east of London was purchased.]

To Catherine Darwin [24 July 1842]

[12 Upper Gower Street]
Sunday

My dear Catty

You must have been surprised at not having heard sooner about the House.. Emma & I only returned yesterday afternoon from sleeping there.— I will give you in detail, as my Father would like, **my** opinion on it.— Emma's slightly differs.— Position.—about $\frac{1}{4}$ of a mile from small village of Down in Kent 16 miles from St. Pauls— eight-miles & $\frac{1}{2}$ from Station, (with many trains) which station is only 10 from London— This is bad, as the drive from the hills is long.— I calculate we are two hour's journey from London Bridge. ... Village about 40 houses with old walnut trees in middle where stands an old flint Church & the lanes meet.— Inhabitants very respectable.— infant school— grown up people great musicians— all touch their hats as in Wales, & sit at their open doors in evening, no high-road leads through village.— The little pot-house, where we slept is a grocers-shop & the land-lord is the carpenter—so you may guess style of village— There are butcher & baker & post-office.— A carrier goes weekly to London & calls anywhere for anything in London, & takes anything anywhere.— On the road to the village, on *fine day* scenery absolutely beautiful: from close to our house, view, very distant & rather beautiful—but house being situated on rather high table-land, has somewhat of desolate air— There is most beautiful old farm-house with great thatched barns & old stumps, of oak-trees like that of Shelton, one field off.— The charm of the place to me is that almost every field is intersected (as alas is our's) by one or more foot-paths— I never saw so many walks in any other country— The country is extraordinarily rural & quiet with narrow lanes & high hedges & hardly any ruts— It is really surprising to think London is only 16 miles off.— The house stands very badly close to a tiny lane & near another man's field— Our field is 15 acres & flat, looking into flat-bottomed valleys on both sides, but no view from drawing-room, wh: faces due South except our own flat field & bits of rather ugly distant horizon.— Close in front, there are some old (very productive) cherry-trees, walnut-trees.—yew.—spanish-chesnut,—pear—old

74

larch, scotch-fir & silver fir & old mulberry-trees make rather a pretty group— They give the ground an old look, but from not flourishing much also give it rather a desolate look. There are quinces & medlars & plums with plenty of fruit, & Morells-cherries, but few apples.— The purple magnolia flowers against house: There is a really fine beech in view in our hedge.— The Kitchen garden is a detestable slip & the soil looks wretched from quantity of chalk flints, but I really believe it is productive. The hedges grow well all round our field, & it is a noted piece of Hay-land This year the crop was bad, but was bought, as it stood, for 2£ per acre, that: is 30£.—the purchaser getting it in— Last year it was sold for £45.—no manure put on in interval. Does not this sound well **ask my father**? Does the mulberry & magnolia show it is not very cold in winter, which I fear is the case.— Tell Susan it is 9 miles from Knole Park—6 from Westerham—seven from Seven-Oaks—at all which places I hear scenery is beautiful.— There are many very odd views round our house deepish flat-bottomed valley & nice farm-house, but big white, many, ugly fallow fields;—much wheat grown here — —

House ugly, looks neither old nor new.—walls two feet thick—windows rather small—lower story rather low.— Capital study 18 × 18. Dining room, 21. × 18.— Drawing-room can easily be added to is 21. × 15. Three stories, plenty of bed-rooms— We could hold the Hensleighs & you & Susan & Erasmus all together.— House in good repair Mʳ Cresy a few years ago laid out for the owner 1500£ and made new roof— Water-pipes over—two bath-rooms—pretty good office & good stable yard & & a cottage.— House in good repair.— I *believe* the price is about 2200£, & I have no doubt I shall get it for one year on lease first to try.—so that I shall do nothing to house at first.—

(Last owner kept 3 cows, one horse & one donkey & sold some hay annually from our field)—. I have no doubt, if we complete purchase, I shall at least save 1000£ over Westcroft, or any other house. we have seen— Emma was at first a good-deal disappointed & at the country round the house; the day was gloomy & cold with NE wind. She likes the actual field & house better than I; the house is just situated, as she likes for retirement, not too near or too far from other houses—but she thinks the country looks desolate— I think all chalk-countries do, but I am used to Cambridgeshire, which is ten times worse.— Emma is rapidly coming round.— she was dreadfully bad with toothache, headache, in the evening, of Friday—but in coming back yesterday she was so delighted with the scenery for the first few miles from Down, that it has worked great change in her.— We go there again the first fine day Emma is able & we then finally settle what to do. . . .

The great Astronomer Sir J. Lubbock is owner of 3000 acres here, & is building a grand house a mile off— I believe he is very reserved & shy & proud or fine—so I suspect he will be no catch, & will never honour us . . .

[Incomplete]

To George Robert Waterhouse   [26 July 1843]

Down, Bromley Kent
Wednesday

My dear Waterhouse

Now for a letter in answer to your two ones on classification—on which I have been often thinking. It has long appeared to me, that the root of the difficulty in settling such questions as yours,—whether number of species &c &c should enter as an element in settling the value or existence of a group—lies in our ignorance of what we are searching after in our natural classifications.— Linnæus confesses profound ignorance.— Most authors say it is an endeavour to discover the laws according to which the Creator has willed to produce organized beings— But what empty high-sounding sentences these are— it does not mean order in time of creation, nor propinquity to any one type, as man.— in fact it means just nothing— According to my opinion, (which I give every one leave to hoot at, like I should have, six years since, hooted at them, for holding like views) classification consists in grouping beings according to their actual *relationship,* ie their consanguinity, or descent from common stocks . . . To me, of course, the difficulty of ascertaining true relationship ie a natural classification remains just the same, though I know what I am looking for. . . .

There is one caution, which should not be overlooked, namely the great doubt whether the groups, which are *now* small, may not have been at some former time abundant: and you will admit fossil & recent beings all come into one system.— In fish, it would appear, that some of the main divisions, which are now least abundant in species, appear formerly to have been most so. It w^d take a Chapter to argue, how probable it is that Geology has never revealed & never will reveal, more than one out of a million forms, which have existed. . . .

Yours ever, C. D.

[Shortly after CD's marriage, Syms Covington, his servant and assistant during and after the *Beagle* voyage, migrated to Australia.]

To Syms Covington   7 October 1843

> Down, near Bromley, Kent
> October 7th 1843
>
> N.B. This will be my direction for the rest of my life.

Dear Covington.

Your new ear trumpet has gone by the ship Sultana. It is enclosed in a box from Messrs Smith & Elder to their correspondent, Mr. Evans (I suppose, bookseller). I was not able to get it sent sooner. You must accept it as a present from me.

I presume you will have to pay a trifle for carriage. I recommend you to take your old one to some skilful tinman, and by the aid of an internal plaster cast I have no doubt he could make them. All that is required is an exact resemblance in form. I should think it would answer for him to make one, & hang it up in his shop with an advertisement.

I was glad to get your last letter with so good an account of yourself, and that you had made a will. My health is better since I have lived in the country.

I have now three children. I am yet at work with the materials collected during the voyage. My coral-reef little book has been published for a year—the subject on which you copied so much M.S. The Zoology of the voyage of the Beagle is also completed.

I have lately heard that the Beagle has arrived safe & sound in the Thames, but I have heard no news of any of the Officers. Your friends at Shrewsbury often enquire after you. I forget whether I ever [told] you that Mrs. Evans is married & that my father has built them a nice little house to live in.

Captain Fitzroy you will have heard, is gone to New Zealand as Governor. I believe he intended to call at Sydney.

With best wishes for your prosperity, which is sure to follow you if you continue in your old, upright, prudent course.

Believe me, yours very faithfully | C. Darwin.

[In September 1843, Joseph Dalton Hooker, a young acquaintance of CD's, returned to England from a four-year survey of Antarctica under James Clark Ross.]

To Joseph Dalton Hooker [13 or 20 November 1843]

Down near Bromley | Kent

Monday

My dear Sir

I had hoped before this time to have had the pleasure of seeing you & congratulating you on your safe return from your long & glorious voyage.[5]

But as I seldom go to London, we may not yet meet for some time—without you are led to attend the Geological Meetings.

I am anxious to know what you intend doing with all your materials— I had so much pleasure in reading parts of some of your letters, that I shall be very sorry if I, as one of the Public, have no opportunity of reading a good deal more.— I suppose you are very busy now & full of enjoyment; how well I remember the happiness of my first few months of England— it was worth all the discomforts of many a gale— But I have run from the subject, which made me write, of expressing my pleasure that Henslow, (as he informed me a few days since by letter) has sent to you my small collection of plants— You cannot think how much pleased I am, as I feared they w^d have been all lost & few as they are, they cost me a good deal of trouble.— There are a very few notes, which I believe Henslow has got describing the habitats &c of some few of the more remarkable plants.— I paid particular attention to the Alpine flowers of Tierra Del. & I am sure I got every plant, which was in flower in Patagonia at the seasons, when we were there.— I have long thought that some general sketch of the Flora of that point of land, stretching so far into the southern seas, would be very curious.— Do make comparative remarks on the species allied to the Europæan species, for the advantage of Botanical Ignoramus'es like myself. It has always struck me as a curious point to find out, whether there are many Europæan genera in T. del Fuego, which are not found along the ridge of the Cordillera; the separation in such cases w^d be so enormous.— Do point out in any sketch you draw up, what genera are American & what Europæan & how great the differences of the species, are, when the genera are Europæan, for the sake of the Ignoramuses.—

I hope Henslow will send you my Galapagos Plants (about which Humboldt even expressed to me considerable curiosity)— I took much pains in collecting all I could,— A Flora of this archipelago would, I suspect, offer a nearly parallel case to that of St Helena, which has so long excited interest.

Pray excuse this long rambling note, & believe me, my dear Sir |
Yours very sincerely | C. Darwin

Will you be so good as to present my respectful compliments to
Sir W. Hooker.

1844

To J. D. Hooker [11 January 1844]

Down. Bromley Kent
Thursday

My dear Sir

I must write to thank you for your last letter; & to tell you how much all your views & facts interest me.— I must be allowed to put my own interpretation on what you say of "not being a good arranger of extended views"—which is, that you do not indulge in the loose speculations so easily started by every smatterer & wandering collector.— I look at a strong tendency to generalize as an entire evil— . . .

Would you kindly observe one little fact for me, whether any species of plant, *peculiar* to any isl^d, as Galapagos, St. Helena or New Zealand, where there are no large quadrupeds, have hooked seeds,—such hooks as if observed here would be thought with justness to be adapted to catch into wool of animals.—

Would you further oblige me some time by informing me (though I forget this will certainly appear in your Antarctic Flora) whether in isl^d like St. Helena, Galapagos, & New Zealand, the number of families & genera are large compared with the number of species, as happens in coral-isl^d, & as I *believe*? in the extreme Arctic land. Certainly this is case with Marine shells in extreme Arctic seas.— Do you suppose the fewness of species in proportion to number of large groups in *Coral-islets.*, is owing to the chance of seeds from all orders, getting drifted to such new spots? as I have supposed.—

Did you collect sea-shells in Kerguelen land, I sh^d like to know their character.? . . .

Besides a general interest about the Southern lands, I have been now ever since my return engaged in a very presumptuous work & which I know no one individual who w^d not say a very foolish one.— I was so struck with distribution of Galapagos organisms &c &c & with the character of the American fossil mammifers, &c &c

that I determined to collect blindly every sort of fact, which c^d bear any way on what are species.— I have read heaps of agricultural & horticultural books, & have never ceased collecting facts— At last gleams of light have come, & I am almost convinced (quite contrary to opinion I started with) that species are not (it is like confessing a murder) immutable. Heaven forfend me from Lamarck nonsense of a "tendency to progression" "adaptations from the slow willing of animals" &c,—but the conclusions I am led to are not widely different from his—though the means of change are wholly so— I think I have found out (here's presumption!) the simple way by which species become exquisitely adapted to various ends.—[1] You will now groan, & think to yourself 'on what a man have I been wasting my time in writing to.'— I sh^d, five years ago, have thought so. . . .

Believe me my dear Sir | Very truly your's | C. Darwin

To J. D. Hooker   23 February [1844]

Down Bromley Kent
Feb. 23^d

Dear Hooker.

I hope you will excuse the freedom of my address, but I feel that as co-circum-wanderers & as fellow labourers (though myself a very weak one) we may throw aside some of the old-world formality. . . . I have just finished a little volume on the volcanic isl^d. which we visited; I do not know how far you care for dry simple geology, but I hope you will let me send you a copy.— I suppose I can send it from London by common coach conveyance. . . .

I am going to ask you some *more* questions, though I daresay, without asking them, I shall see answers in your work, when published, which will be quite time enough for my purposes. First for the Galapagos, you will see in my Journal, that the Birds, though peculiar species, have a most obvious S. American aspect: I have just ascertained the same thing holds good with the sea-shells.— Is it so with those plants, which are peculiar to this archipelago; you state that their numerical proportions are continental (is not this a very curious fact?) but are they related in forms to S. America.— Do you know any other cases of an Archipelago, with the separate islands possessing distinct representative species? . . .

I hope you will try to grudge as little as you can the trouble of my letters, & pray believe me, very truly your's, | C. Darwin

[By 1844, CD had made only one attempt to set down any theoretical conclusions that he felt were justified by the data he had collected. This was a rough pencil sketch of 35 pages, written in 1842. Two years later, he wrote a longer, more carefully composed manuscript of 231 pages that, in the event of his death, he felt could be the basis for a publication of his theory.]

To Emma Darwin   5 July 1844

Down.

July 5th.— 1844

My. Dear. Emma.

I have just finished my sketch of my species theory. If, as I believe that my theory is true & if it be accepted even by one competent judge, it will be a considerable step in science.

I therefore write this, in case of my sudden death, as my most solemn & last request, which I am sure you will consider the same as if legally entered in my will, that you will devote 400£ to its publication & further will yourself, or through Hensleigh, take trouble in promoting it.— I wish that my sketch be given to some competent person, with this sum to induce him to take trouble in its improvement. & enlargement.— I give to him all my Books on Natural History, which are either scored or have references at end to the pages, begging him carefully to look over & consider such passages, as actually bearing or by possibility bearing on this subject.— I wish you to make a list of all such books, as some temptation to an Editor. I also request that you hand over him all those scraps roughly divided in eight or ten brown paper Portfolios:— The scraps with copied quotations from various works are those which may aid my Editor.— I also request that you (or some amanuensis) will aid in deciphering any of the scraps which the Editor may think possibly of use.— I leave to the Editor's judgment whether to interpolate these facts in the text, or as notes, or under appendices. As the looking over the references & scraps will be a long labour, & as the **correcting** & enlarging & altering my sketch will also take considerable time, I leave this sum of 400£ as some remuneration & any profits from the work.— I consider that for this the Editor is bound to get the sketch published either at a Publishers or his own risk. Many of the scraps in the Portfolios contains mere rude suggestions & early views now useless, & many of the facts will probably turn out as having no bearing on my theory.

With respect to Editors.— M^r. Lyell would be the best if he would undertake it: I believe he w^d find the work pleasant & he w^d learn some facts new to him. As the Editor must be a geologist, as well as Naturalist. The next best Editor would be Professor Forbes of London. The next best (& quite best in many respects) would be Professor *Henslow*??. D^r. Hooker would perhaps correct the Botanical Part probably—he would do as Editor— Dr Hooker would be **very** good The next, Mr Strickland.— If no⟨ne⟩ of these would undertake it, I would request you to consult with Mr Lyell, or some other capable man, for some Editor, a geologist & naturalist. ...

My dear Wife | Yours affect | C. R. Darwin

If there sh^d be any difficulty in getting an editor who would go thoroughily into the subject & think of the bearing of the passages marked in the Books & copied out on scraps of Paper, then let my sketch be published as it is, stating that it was done several years ago & from memory, without consulting any works & with no intention of publication in its present form ...

To Leonard Horner   29 August [1844]

Down near Bromley | Kent
Aug 29th.

My dear M^r. Horner

... I have been lately reading with care A. d'Orbigny work on S. America, & I cannot say how forcibly impressed I am with the infinite superiority of the Lyellian school of Geology over the Continental. I always feel as if my books came half out of Lyell's brains & that I never acknowledge this sufficiently, nor do I know how I can, without saying so in so many words—for I have always thought that the great merit of the Principles,^{2} was that it altered the whole tone of one's mind & therefore that when seeing a thing never seen by Lyell, one yet saw it partially through his eyes— it would have been in some respects better if I had done this less—but again excuse my long & perhaps you will think presumptuous discussion. ...

Believe me | dear Mr Horner | Yours truly obliged | Charles Darwin

To J. D. Hooker [8 September 1844]

Down Bromley Kent
Sunday

My dear Hooker

... The subject of the greater number of species in certain areas, than in others, has long appeared to me a very curious subject: your example of East S. America, compared with Britain is very striking. Is not the case of New Zealand, with its varied stations, compared with the uniformly arid C. of Good Hope, opposed to your view, that the number of species bears a relation to the vicissitudes of climate? When you speak of mountains, (as the plains of the Andes) being subject to vicissitudes, I am not sure, whether you means absolutely so on the same spot, or whether great differences, within short distances.— Is it not said, that the absolute changes of temperature are greatest on any one spot, in the extreme northern regions; & that equability is the characteristic of the tropics?

The conclusion, which I have come at is, that those areas, in which species are most numerous, have oftenest been divided & isolated from other areas,, united & again divided;—a process implying antiquity & some changes in the external conditions. This will justly sound very hypothetical.

I cannot give my reasons in detail: but the most general conclusion, which the geographical distribution of all organic beings, appears to me to indicate, is that isolation is the chief concomitant or cause of the appearance of *new* forms (I well know there are some staring exceptions).—

Secondly from seeing how often plants & animals swarm in a country, when introduced into it, & from seeing what a vast number of plants will live, for instance in England, if kept *free from weeds & native plants*, I have been led to consider that the spreading & number of the organic beings of any country depend less on its external features, than on the number of forms, which have been there originally created or produced.— I much doubt whether you will find it possible to explain the number of forms by proportional differences of exposure; & I cannot doubt if half the species in any country were destroyed or had not been created, yet that country w^d: appear to us fully peopled. With respect to original creation or production of new forms, I have said, that isolation appears the chief element: Hence, with respect to terrestrial productions, a tract of country, which had oftenest within the later geological periods subsided & been converted into isl^{ds}, & reunited, I sh^d expect to contain most forms.—

But such speculations are amusing only to one self, & in this case useless as they do not show any direct line of observation: if I had foreseen how hypothetical, the little, which I have *unclearly* written, I w^d not have troubled you with the reading of it.

Believe me,—at last not hypothetically— | Yours very sincerely | C. Darwin . . .

84

To Leonard Jenyns   12 October [1844]

Down Bromley Kent
Oct 12th

My dear Jenyns

... I am surprised that with all your parish affairs that you have had time to do all, that which you have done. I shall be very glad to see your little work[3] (& proud sh^{d}. I have been, if I could have added a single fact to it): my work on the species question has impressed me very forcibly with the importance of all such works, as your intended one, containing what people are pleased generally to call trifling facts. These are the facts, which make one understand the working or œconomy of nature. There is one subject, on which I am very curious, & which perhaps you may throw some light on, if you have ever thought on it,—namely what are the checks & what the periods of life, by which the increase of any given species is limited. Just calculate the increase of any bird, if you assume that only half the young are reared & these breed: within the *natural* ie. if free from accidents life of the parents, the number of individuals will become enormous, & I have been much surprised to think, how great destruction *must* annually or occasionally be falling on every species, yet the means & period of such destruction scarcely perceived by us.

I have continued steadily reading & collecting facts on variation of domestic animals & plants & on the question of what are species; I have a grand body of facts & I think I can draw some sound conclusions. The general conclusion at which I have slowly been driven from a directly opposite conviction is that species are mutable & that allied species are co-descendants of common stocks. I know how much I open myself, to reproach, for such a conclusion, but I have at least honestly & deliberately come to it.

I shall not publish on this subject for several years— At present I am on the geology of S. America. I hope to pick up from your book, some facts on slight variations in structure or instincts in the animals of your acquaintance

Believe me Ever yours | C. Darwin

To Henry Denny   7 November [1844]

Down near Bromley | Kent
Nov. 7

Dear Sir

I am much obliged for your note & have been greatly interested by the facts you mention of the identical parasites on the same species of

85

birds at immensely remote stations. I am sorry to say I cannot think of any possible means of procuring the parasites of the S. American Mammifers to which you refer. Some surgeon, or officer, interested in Nat. Hist w^d be the only means & I know none now there.

Are you aware whether the same parasites are found on any of our *land* birds in this country & in N. America. Some of the birds of Europe & N. America appear certainly identical; many form very closely related species or as some would think races: What an *interesting* investigation w^d be the comparison of the parasites of the closely allied & representative birds of the two countries.

Should you chance to know anything of the parasites of the *land*-birds of North America, perhaps, sometime you kindly w^d take the trouble to send me a line, as I am deeply interested in everything connected with geographical distribution, & the differences between species & varieties.— I hope you will turn in your mind the possibility of investigating closely the N. American land-bird-parasites.—

When the same bird in *immensely* remote countries, has the same parasite, do you never observe some slight difference in colour, size or proportions of such parasites? I have forgotten to answer your question, about the Aperea being identical with the guinea-pig, & I can only answer it by professing entire ignorance & doubt: I certainly sh^d disbelieve it, if you c^d show the parasites were different. How is the parasite of the wolf with the dog, if the latter has one?

I hope you will excuse this long note & believe me dear Sir | Yours very faithfully | C. Darwin

1845–1846

[Down]
Monday night

My dear Wife

Now for my day's annals— In the morning I was baddish, & did hardly any work & was as much overcome by my children, as ever Bishop Coplestone was with Duck. But the children have been very good all day, & I have grown a good deal better this afternoon, & had a good romp with Baby—[1] I see, however, very little of the Blesseds— The day was so thick & wet a fog, that none of them went out, though a thaw & not very cold; I had a long pace in the Kitchen Garden: Lewis came up to mend the pipe & paper the W.C. in which apartment there was a considerable crowd for about an hour, when Mr Lewis & his son William, Willy Annie, Baby & Bessy were there. Baby insisted on going in, I daresay, greatly to the disturbance of Bessy's delecacy— Lewis from first dinner to second dinner was a first-rate dispensary, as they never left him— They, also, dined in the Kitchen, and I believe have had a particularly pleasant day.—

I was playing with Baby in the window of the drawing-room this morning, & she was blowing a feeble fly (fry) & blew it on its back, when it kicked so hard, that to my great amusement Baby grew red in the face, looked frightened & pushed away from the window.— The children are growing so quite out of all rule in the drawing-room, jumping on everything & butting like young bulls at every chair & sofa, that I am going to have the dining-room fire lighted tomorrow & keep them out of the drawing-room. I declare a months such wear, w^d spoil every thing in the whole drawing-room.—

I read Whately's Shakspeare & very ingenious & interesting it is—and what do you think Mitford's Greece has made me begin, the Iliad by Cowper, which we were talking of; & have read 3 books with much more pleasure, than I anticipated.— I have given up acids & gone to puddings again.—

Tuesday morning— I am impatient for your letter this morning to hear how you got on.— I asked Willy how Baby has slept & he answered "she did not cry not one mouthful". My stomach is baddish again this morning & I almost doubt, whether I will go to London, tomorrow; if I do you won't hear. Poor Annie has had a baddish knock by Willie's ball in her eye.— it is swelled a bit, but not otherwise bad.

    C. D. ...

To J. D. Hooker [11–12 July 1845]

Down Bromley Kent

Friday

My dear Hooker

I sh^d have written to you a few days ago, as I had some questions to ask & several points in your last letter, which I should much enjoy discussing with you: but on Wednesday an upsetting event happened in the fact of a Boy-Baby being born to us— may he turn out a Naturalist.[2] My wife is going on most comfortably.— First I have got a few questions about the Galapagos Plants, as I am now come (not in correcting press, but first time over) to that Chapter:[3] I will put these questions on a separate paper & some of them you can answer by a word or two on the paper on its back & return it to me, pretty soon, if you can so manage it. I cannot tell you how delighted & astonished I am at the results of your examination; how wonderfully they support my assertion on the differences in the animals of the different islands, about which I have always been fearful: I see the case excites the interest even of R. Brown. ...

Have you any good evidence for absence of insects in small islands: I found 13 species in Keeling atoll. Flies are good fertilisers; & I have seen a microscopic Thrips & a Cecidomya take flight from a flower in the direction of another with pollen adhering to them. In Arctic countries a Bee seems to go as far N. as any flower.— Not that I am a Believer in Hybridising to any extent worth mention; but I believe the absence of insects w^d present the most serious difficulty to the impregnation of a host of (not diœcious or monœcious plant) plants ...

Without knowing the age of the Kerguelen tree no one would, I presume, guess about any change of climate since they grew: S. America was once hotter, then much colder, than now: in N. America, within Tertiary epochs the series, has been.—hot—warm—very cold—a little warmer than now—present climate. ...

I am not aware that I want any geological information from India; but if your friend resides near those parts where the Chetah is used for hunting I am *particularly* anxious to know, whether they *ever* breed in domestication; & if never or seldom, whether they copulate, & whether it is thought to be the fault of the male or female.— Again if he reside in the silk-worm districts, any information whether the moths, caterpillars or cocoons vary at all,—whether the inhabitants take any pains in selecting good individuals for breeding—whether there is any traditional belief in the origin of any breed, ie if different breeds of the same species are found in different districts.—or any analagous information.— This w^d. be eminently valuable to me. . . .
    Ever yours | C. Darwin . . .

To J. D. Hooker [10 September 1845]
Down Bromley Kent
Wednesday
My dear Hooker
    . . . Many thanks for your letter received yesterday, which, as always, sets me thinking: I laughed at your attack at my stinginesss in changes of level towards Forbes, being so liberal towards myself; but I must maintain, that I have never let down or upheaved our mother earth's surface, for the sake of explaining any one phenomenon, & I trust I have very seldom done so without some distinct evidence. So I must still think it a bold step, (perhaps a very true one) to sink into depths of *ocean, within the period of* **existing species**, so large a tract of surface. But there is no amount or extent of change of level, which I am not fully prepared to admit, but I must say I sh^d. like better evidence, than the identity of a few plants, which *possibly* (I do not say probably) might have been otherwise transported. . . .
    How painfully (to me) true is your remark that no one has hardly a right to examine the question of species who has not minutely described many. I was, however, pleased to hear from Owen (who is vehemently opposed to any mutability in species) that he thought it was a very fair subject & that there was a mass of facts to be brought to bear on the question, not hitherto collected. My only comfort is, (as I mean to attempt the subject) that I have dabbled in several branches of Nat. Hist: & seen good specific men work out my species & know something of geology; (an indispensable union) & though I shall get more kicks than half-pennies, I will, life serving, attempt my work.— Lamarck is the only exception, that I can think

of, of an accurate describer of species at least in the Invertebrate kingdom, who has disbelieved in permanent species, but he in his absurd though clever work has done the subject harm, as has M^r. Vestiges,[4] and, as (some future loose naturalist attempting the same speculations will perhaps say) has M^r. D. ...

To J. D. Hooker   [3 September 1846]

Down Farnborough Kent
Thursday

My dear Hooker

I hope this letter will catch you at Clifton, but I have been prevented writing by being unwell & having had the Horner's here as visitor, which with my abominable press-work has fully occupied my time. It is, indeed, a long time since we wrote to each other; though, I beg to tell you, that I wrote last, but what about I cannot remember, except, I know, it was after reading your last numbers, & I sent you a uniquely laudatory epistle, considering that it was from a man who hardly knows a daisy from a Dandelion to a professed Botanist. ...

I was very glad to hear what you were about; but I fear you must feel your time rather thrown away.— I cannot remember, what papers have given me the impression, but I have that, which you state to be the case, firmly fixed on my mind, namely the little chemical importance of the soil to its vegetation.— What a strong fact it is, as R Brown once remarked to me, of certain plants being calcareous ones here which are not so under a more favourable climate on the continent, or the reverse, for I forget which; but you no doubt will know to what I refer.— By the way there are some such cases in Herbert's paper in Hort. Journal:[5] have you read it, it struck me as extremely original & bears *directly* on your present researches.— To a *non-botanist* the Chalk has the most peculiar aspect of any flora in England; why will you not come here & make your observations? *We* go to Southampton, if my courage & stomach do not fail, for the Brit. Assoc: (do you not consider it your duty to be there?), & why cannot you come here afterwards & *work*. I expect Sulivan here the first week in October & I hope to get a few more here, & how glad we should be if you c^d come then or at any time whatever.—

Before the end of the month, I shall have quite finished my S. American Geology,[6] & extremely glad I shall be, for I have been pushing on & feeling jaded for the last several months by it.—

I am astonished (having felt a curiosity on the point) at the number of species on 2 square yards (or two yards square?); though I cannot read whether it is *26* or *16* to *48* species; does this include cryptogams: if you do *not* publish this, I sh^d like much sometime to hear more particulars about this; if you publish, where will it be?

I am much pleased to hear you have worked out the identical & representative species of N. temperate & Antarctic regions & shall be exceedingly glad to see it;[7] but as it **of course** will be published, I will not think of troubling you to send it me: I hope you will add, whenever you know, whether species of the same genera are found in the intermediate *tropical* districts, saying, whether in America or elsewhere, whether on high-lands or lowlands; this no doubt w^d add to your trouble, even if you gave *only* such information as you possessed without search, & surely it w^d add great interest to your results . . .

Have you ever thought of G. St. Hilaire "loi du balancement", as applied to plants: I am well aware that some zoologists quite reject it, but it certainly appears to me, that it often holds good with animals.— You are no doubt aware of the kind facts I refer to, such as great development of canines in the carnivora apparently causing a diminution—a compensation or balancement—in the small size of premolars &c &c.— I have incidentally noticed some analogous remarks on plants, but have never seen it discussed by Botanists.— Can you think of cases of any one species in genus, or genus in family, with certain parts extra developed, & some adjoining parts reduced? In varieties of same species, double flowers & large fruits seem something of this,—want of pollen & of seeds balancing with the increased number of petals & development of fruit. . . .

Ever my dear Hooker | Most truly yours | C. Darwin . . .

1847

To J. D. Hooker [1 May 1847]

Down Farnborough Kent.

Saturday

My dear Hooker

I send the accompanying pamphlet[1] (which may be left *anytime* at Athenæum or Geolog Soc.) for the *chance* of your not having seen it & your liking to do so.— The Geological reasoning appears to me quite sound, except touching the old shallow seas. I am delighted to hear that Brongiart thought Sigillaria aquatic & that Binney considers coal a sort of submarine peat. I w^d bet 5 to 1 that in 20 years this will be generally admitted; and I do not care for whatever the Botanical difficulties or impossibilities may be. If I could but persuade myself that Sigillaria & Co. had a good range of depth, ie c^d live from 5 to 100 fathoms under water, all difficulties of nearly all kinds would be removed.—(for the simple fact of muddy ordinarily shallow sea implies proximity of land.) (NB I am chuckling to think how you are sneering all this time.) It is not much of a difficulty there not being shells with the coal, considering how unfavourable deep mud is for most Mollusca: & that shells w^d probably decay from the humic acid, as seems to take place in peat & in the *black* moulds (as Lyell tells me) of the Missisippi.— so coal question settled. Q.E.D— sneer away.—

Many thanks for your welcome note from Cambridge & I am glad you like my alma mater, which I despise heartily as a place of education, but love from many most pleasant recollections; I am delighted to think there is any chance of Henslow & you coming here; you did very right to urge him here.— I hope much to be at Oxford, but my poor wife will be otherwise engaged[2] & that is my only cause of doubt of being able to attend. ...

Farewell my dear Hooker & be a good boy & make Sigillaria a submarine sea-weed ... Ever yours. C. Darwin

92

['[*Hooker*] *is in all ways very impulsive and somewhat peppery in temper; but the clouds pass away almost immediately. He once sent me an almost savage letter from a cause which will appear ludicrously small to an outsider, viz. because I maintained for a time the silly notion that our coal-plants had lived in shallow water in the sea.' (Autobiography*, p. 105.)]

To J. D. Hooker   [6 May 1847]

Down

Thursday

My dear Hooker

You have made a savage onslaught, & I must try to defend myself. But first let me say that I never write to you except for my own good pleasure; now I fear that you answer me when busy & without inclination (& I am sure I sh^d. have none, if I was as busy as you): pray do not do so, as if I thought my writing entailed an answer from you nolens volens, it would destroy all my pleasure in writing.—

Firstly: I did not consider my letter as *reasoning*, or even as *speculation*, but simply as mental rioting & as I was sending Binney's paper I poured out to you the result of reading it.—

Secondly, you are right indeed in thinking me mad, if you suppose that I would class any ferns as marine plants: but surely there is a wide distinction between the plants found upright in the coal beds & those not upright & which might have been drifted. Is it not probable that the same circumstances which have preserved the vegetation *in situ*, sh^d. have preserved drifted plants? I know calamites is found upright, but I fancied its affinities were very obscure like Sigillaria. As for Lepidodendron I forgot its existence, as happens when one goes riot & now know neither what it is, or whether upright. If these plants, ie calamites & Lepidodendron have *very clear relations* to terrestrial vegetables,, like the ferns have, & are found upright in situ, of course I must give up the ghost. But surely Sigilliria is the main upright plant, & on its obscure affinities I have heard you enlarge.—

Thirdly, it never entered my head to under=value botanical relatively to zoological evidence; except in so far as I thought it was admitted that the vegetative structure seldom yielded any evidence of affinity, nearer than that of families, & not always so much: & is it not in plants, as certainly it is in animals, dangerous to judge of habits without very near affinity. Could a Botanist tell from structure alone that the mangrove family, almost or quite alone in dicots:, could live in the sea—& the zostera family almost alone amongst the monocot^s:? Is it a safe argument, that because algæ are almost the only, or the only, submerged sea-plants, that formerly other groups had not members with such habits; with animals such an argument would not be conclusive, as I c^{d} illustrate by many examples;—but I am forgetting myself, I want only to some degree to defend myself, & not burn my fingers by attacking you.— The foundation of my letter, & what is my deliberate opinion, though I daresay you will think it absurd, is that I would rather trust, cæteris paribus, pure geological evidence than either Zoolog. or Botan. evidence: I do not say that I w^{d} sooner trust *poor* geolog. evidence than *good* organic: I think the bases of pure geological reasoning is simpler, (consisting chiefly of the action of water on the crust of the earth, & its up & down movements) than bases drawn from the difficult subject of affinities & of structure in relation to habits. . . .

Tell me that an upright fern *in situ* occurs with Sigilliria & Stigmaria, or that the affinities of Calamites & Lepidodendron (supposing that they are found in situ with Sigilliria) as so *clear* that they could not have been marine, like, but in a greater degree, than the mangrove & sea-wrack, & I will humbly apologise to you & all Botanists, for having let my mind run riot on a subject on which assuredly I know nothing. But till I hear this, I shall keep privately to my own opinion. . . .

Whether this letter will sink me still lower in your opinion, or put me a little right, I know not, but hope the latter. Anyhow I have revenged myself with boring you with a very long epistle.

Farewell & be forgiving— Ever yours C. Darwin . . .

To George Grey   13 November 1847
Down, Farnborough Kent.

Nov. 13/47/

My dear Sir

. . . I am extremely glad to know how well your Colony is now prospering. Ever since the voyage of the Beagle, I have felt the deepest interest with respect to all our colonies in the southern hemisphere. However much trouble & anxiety you must have had & will still have, it must ever be the highest gratification to you to reflect on the prominent part you have played in two countries, destined in future centuries to be great fields of civilization.—

You are so kind as to offer aid in any Natural History researches in New Zealand: I have no *personal* interest on any point there; but there are two subjects which have long appeared to me well deserving investigation; & if hereafter your labours should be lightened you might like to attend to them yourself, or direct the attention to them of any Naturalist under you.— The first is, an examination of any limestone caverns: such exist near the Bay of Islands & I daresay elsewhere: I was prevented entering them by their having been used as places of Burial. Digging in the mud under the usual stalagmitic crust, would probably reveal bones of the cotemporaries of the Dinornis. I think there is a special interest on this point, from New Zealand being at present so eminently instructive in a negative point of view, with respect to the distribution of terrestrial mammifers. I have long ardently wished to hear of the exploration of the New Zealand Caverns.

The second point is, whether there are "erratic boulders" in New Zealand, more especially in the middle & southern islands, & their northern limit, if such occur. Most geologists are now united to considering erratic boulders, to have been transported by icebergs or glaciers. I consider it as a most important question, *as bearing upon the former climate* of the world, to know whether such proofs occur generally in the S. hemisphere as in the Northern: I have ascertained that such is the case in S. America, from Cape Horn to about Lat 40°. This subject requires much care & some little knowledge or at least thought. I saw inland of the Bay of Islands, large rounded blocks of greenstone, but I was unable to ascertain whether the parent rock was far distant; nor did I then see the full importance of the question, otherwise I would have devoted every hour to it. Very large, *angular* blocks of foreign rock, lying on *isolated* hills or hillocks offer the best & without much care, the only sure evidence. We do not know the extent to which during ages, the waves of the sea, at various alternating levels, with earthquake waves &c & occasional heavy floods, may transport in valleys & over an undulating surface very large boulders, hence becoming rounded. Granite from its tendency to orbicular disintegration has given rise to several false accounts of erratic boulders.— . . .

If I have wearied you with these details, I beg to apologise & you can burn this letter; but I thought, perhaps, you would not object to hear my opinion on two, as I believe, really curious subjects for investigation. I would, myself, go through much labour to investigate the erratic phenomena & trace its limits & age. Should you ever obtain any evidence on this head, it would delight me to hear the result.

Again allow me to thank you very cordially & I beg your Excellency to believe me | Your sincerely obliged | C. Darwin

To His Excellency | Sir G Grey

To John Edward Gray   18 December 1847

Down Farnborough Kent
December 18th 1847

My dear Gray

You are aware that I have been attending for the last 14 months to the anatomy of the various genera of Cirripedia. Having, as I hope, now acquired a fair knowledge of their fundamental structure, it is my intention to publish a monograph on this difficult order. The object of this letter is to ask you to request the permission of the Trustees to describe the Public collection of the Museum. This, however, involves the absolute necessity of my having the collection, not all at once, but in groups at my house here. I find by experience that each species takes me between 2 & 3 days, & each new genus, as many weeks. Every portion requires examination under the microscope & all the minuter organs under a high compound power: the shells also, require soaking & cleaning. I have resolved not to describe any species, without I can do it thoroughily. I am well aware that my request in an unusual one; but I would most respectfully beg to call the attention of the Trustees to the fact that specimens are sent out to be mounted, & that one specimen of every species of Cirripede must be disarticulated for the characters to be ascertained, & the parts of the mouth dissected. The portions thus dissected I prepare in spirits between two plates of glass. If the Trustees think me worthy of their confidence I will give to the Museum all such preparations, (whether made from my own or the Public collection) & all my entire shells (including many new species), as soon as my work is completed. I would further beg to call the attention of the Trustees to the fact, that their entire collection, (contained in 8 or 10 drawers) will thus be named & arranged without the loss of the valuable time of the Officers: though I fully believe that you could do the work in half the time I could, yet I am convinced that to examine & classify the public collection in the Order, as it should be done, would take a year.

In case the Trustees are inclined to do me the honour of acceding to my request, I pledge myself to take the utmost care of the

Collection & to do nothing whatever to the specimens, without your express permission.— ...

How much a monograph of this Order is wanted, you, who know it far better than any man in England, are well aware. In fact the whole of the species are in almost a compte state of chaos: as Agassiz has remarked, a "Monograph of the Cirripedia is now a pressing desideratum in Zoology". How far I am capable of this undertaking you must decide; if I fail it shall not be for want of labour.

I apologise for the length of this letter, & beg to thank you for the kind assistance you have already given me.

I remain | Yours very faithfully | Charles R. Darwin

To J. E. Gray Esq^e

1848

To James Smith   28 January [1848]

Down Farnborough Kent
Jan 28th

My dear Sir

I hope that you will excuse the liberty I take in asking you a great favour. I have been employed for the last year & shall be for, I suppose, the next two years on a Monograph, anatomical & systematic on the whole class of Cirripedia. In the last number of the Geolog. Journal I see that you found in Portugal at least six species of Balanus. Will you entrust me with your specimens to describe, that is if I find I can make any hand at the fossil species of which I have already got some, Mr Lyell, & Mr Stutchbury having placed their collections at my disposal. I shall, however, require to keep them sometime & it is absolutely indispensable for me to break up or make section of at least one specimen of each. The characters, hitherto generally used from external aspect, I find, are usually quite valueless; & the internal structure of shell must be in each case examined. In some genera, as far as I yet see, the included animal alone offers distinguishable characters; so that I have some fears about the fossil species, but I mean to try at them.

My friends have been most generous in placing collections at my disposal. Mr Stutchbury has sent me the whole of his magnificent collection & Mr Cuming has placed his at my disposal. Did you collect any recent species on the coast of Portugal & especially at Madeira; if you have any from these quarters or elsewhere (especially if in Spirits) & would entrust them to me, I should feel very grateful.—

I wish to see the species from as many quarters as possible on account of their Geographical Range. I sh^d be much obliged for any information, on habits frequency &c depths, abundance at bottom of sea in dead state &c, which you would kindly take the trouble to supply me with.

Pray believe me, my dear Sir | Yours very faithfully | C. Darwin

I must express to you, how delighted I was with y⟨ou⟩r paper on Malta;[1] if I was asked for the most striking fact ever discovered, exhibiting the changes of level between land & water & the power of denudation, I should certainly refer to your old roads leading under the sea & over the brink of precipices: such facts seize the imagination with astonishment. It is my belief that if you were confined a prisoner in a Square in London, you would find some demonstrative proof of the level having changed!

To J. S. Henslow   [1 April 1848]
Down Farnborough Kent
Saturday night.

My dear Henslow

... Thanks, ... for your Address,[2] which I like very much. ... I rather demur to one sentence of yours, viz "however delightful any scientific pursuit may be, yet if it shall be wholly unapplied it is of no more use than building castles in the air". Would not your hearers infer from this that the practical use of each scientific discovery ought to be immediate & obvious to make it worthy of admiration? What a beautiful instance Chloriform is of a discovery made from *purely* scientific researches, afterwards coming almost by chance into practical use. For myself I would, however, take higher ground, for I believe there exists, & I feel within me, an instinct for truth, or knowledge or discovery, of something same nature as the instinct of virtue, & that our having such an instinct is reason enough for scientific researches without any practical results *ever* ensuing from them.— You will wonder what makes me run on so, but I have been working very hard for the last 18 months on the anatomy &c of the Cirripedia (on which I shall publish a monograph) & some of my friends laugh at me, & I fear the study of the cirripedia will ever remain "wholly unapplied" & yet I feel that such study is better than castle-building.

Talking of Cirripedia, I must tell you a curious case I have just these few last days made out: all the Cirripedia are bisexual [herm-aphrodite], except one genus, & in this the female has the ordinary appearance, whereas the male has no one part of its body like the female & is microscopically minute; but here comes the odd fact, the male or sometimes two males, at the instant they cease being locomotive larvæ become parasitic within the sack of the female, & thus fixed & half embedded in the flesh of their wives they pass their whole lives & can never move again. Is it not strange that nature should have made this one genus unisexual, & yet have fixed the males on the outside of the females;—the male organs in fact being thus external instead of internal.—³ ...

We are all well here, & a sixth little (d) expected this summer:⁴ as for myself, however, I have had more unwellness than usual.

Believe me, my dear Henslow. | Ever most truly yours | C. Darwin

If you are ever starting any young naturalist with his tools, recommend him to go to Smith & Beck of 6 Colman St. City for a simple microscope: he has lately made one for me, partly from my own model & with hints from Hooker, *wonderfully* superior for coarse and fine dissections than any I ever before worked with. If I had had it sooner, it would have saved me many an hour.—

[Hooker had left on a botanical expedition to India in November 1847. He returned to England on 26 March 1851.]

To J. D. Hooker   10 May 1848

Down Farnborough Kent
May 10/48/

My dear Hooker

I was indeed delighted to see your hand-writing ...

Your letter was the very one to charm me, with all its facts for my species-Book, & truly obliged, I am, for so kind a remembrance of me. Do not forget to make enquiries about origin, even if only traditionally known of any vars. of domestic quadrupeds, birds, **silkworms** &c.— (Are there domestic Bees? if so hive ought to brough home.)

Of all the facts you mention, that of the wild Bhil, when breeding with the domestic, producing offspring, somewhat sterile, is the most surprising; surely they must be different species. Most zoologists would absolutely disbelieve such a statement & consider the result as a *proof* that they were distinct species: I do not go so far as that, but the case seems highly improbable: Blyth has studied the Indian Ruminantia.—

I have been much struck about what you say of lowland plants asending mountains, but the Alpine not descending. How I do hope you will get up some mountains in Borneo; how curious the result will be. By the way I never heard from you, what affinity the Maldiva flora has, which is cruel, as you tempted me by making me guess. I sometimes groan over your Indian Journey, when I think over all your locked up riches: when shall I see a memoir on insular Floras,

& on the Pacific. What a grand subject, Alpine Floras of the world would be, as far as known: and then you have never given a coup d'œil on the similarity & dissimilarity of Arctic & Antarctic floras. Well thank Heavens, when you do come back, you will be nolens-volens a fixture.— I am particularly glad you have been at the Coal: I have often since you went gone on maundering on the subject, & I shall never rest easy in Down church-yard, without the problem be solved by someone before I die.

Talking of dying makes me tell you that my confounded stomach is much the same; indeed of late has been rather worse, but for the last year, I think, I have been able to do more work. I have done nothing besides the Barnacles, except indeed a little theoretical paper on Erratic Boulders, & Scientific Geological Instructions for the Admiralty Volume, which cost me some trouble. This work, which is edited by Sir J. Herschel is a very good job, in as much as, the Captains of Men of War, will now see that the Admiralty care for science & so will favour naturalists on board. As for a man, who is not scientific by nature, I do not believe Instructions will do him any good; & if he be scientific & good for anything the Instructions will be superfluous: I do not know who does the Botany; Owen does the zoology & I have sent him an account of my new simple microscope, which I consider perfect, even better than your's of Chevalier's. N.B. I have got a 18 object glass, & it is grand.— I have been getting on well with my beloved cirripedia, & got more skilful in dissection: I have worked out the nervous system pretty well in several genera, & made out their ears & nostrils, which were quite unknown. I have lately got a bisexual cirripede, the male being microscopically small & parasitic within the sack of the female; I tell you this to boast of my species theory, for the nearest & closely allied genus to it is, as usual, hermaphrodite, but I had observed some minute parasites adhering to it, & these parasites, I now can show, are supplemental males, the male organs in the hermaphrodite being unusually small, though perfect & containing zoosperms: so we have almost a polygamous animal, simple females alone being wanting. I never sh^d. have made this out, had not my species theory convinced me, that an hermaphrodite species must pass into a bisexual species by insensibly small stages, & here we have it, for the male organs in the hermaphrodite are beginning to fail, & independent males ready formed. But I can hardly explain what I mean, & you will perhaps wish my Barnacles & Species theory al Diabolo together. But I don't care what you say, my species theory is all gospel.—

We have had only one party here viz of the Lyells, Forbes, Owen & Ramsay, & we both missed you & Falconer very much. I do not much think we shall have another for my poor dear wife will be employed in July in bringing into the world, under the influence of Chloriform, a sixth little (d) as Henslow calls my children. I know more of your history than you will suppose, for Miss Henslow most goodnaturedly sent me a packet of your letters, & she wrote me so nice a little note that it made me quite proud.— ...

There was a short time since, a not very creditable discussion at meeting of Royal Soc. where Owen fell foul of Mantell with fury & contempt about Belemnites. What wretched doings come from the ardor of fame; the love of truth alone would never make one man attack another bitterly. . . .

pray do not work too hard, my dear Hooker, Your affectionate friend | C. Darwin.—

[From 17 May to 1 June 1848, Darwin visited his ailing father at the family home in Shrewsbury.]

To Emma Darwin   [20–1 May 1848]

[Shrewsbury]
Saturday

My dear Mammy

Though this will not go today, I will write a bit of Journal, which "in point of fact" is a Journal of all our healths. My Father kept pretty well all yesterday, but was able to talk for not more than 10 minutes at a time till after dinner when he talked the whole evening most wonderfully well & cheerfully. It is an inexpressible pleaseure, that he has twice told me that he is very comfortable, & that his want of breath does not distress him at all like the dyeing sensation, which he now very seldom has. That he thought with care he might live a good time longer, & that when he dyed, it would probably be suddenly which was best. Thrice over he has said that he was very comfortable, which was so much more than I expected. Catherine has been having wretched nights, but her spirits appear *to me* as good as used formerly to be.— Lastly for myself, I was a little sick yesterday, but upon the whole very comfortable & I had a splendid good night & am extraordinary well today.—

Thanks for your very nice letter received this morning with all the news about the dear children: I suppose now & be hanged to you, you will allow Annie is something. I believe, as Sir J[ohn W]. L[ubbock] said of his friend, that she a second Mozart, any how she is more than a Mozart, considering her Darwin blood. I am very much puzzled what Annies inciting incident can be. This morning much rain, wind & cold. I have great fears we shall not have our Clock, for I think my Father likes it.— Farewell for today.—

Sunday All goes on flourishing, though I was sick last night, but not very bad. . . .

What a very good girl you are to write me such very nice letters, telling me all I like to hear; though you have not mentioned the 2 new Azaleas.—

Hensleigh thinks he has settled the Free Will question, but here-detariness practically demonstrates, that we have none whatever. . . . I daresay not a word of this note is really mine; it is all hereditary, except my love for you, which I sh^d think could not be so, but who knows?

Yours | C. D.

You were quite right to send the Barnacles; but mind that in all ordinary cases, they must instantly be put in **spirits**.

To J. S. Henslow   2 July [1848]

Down Farnborough Kent
July 2^d.

My dear Henslow

. . . Thanks for your Syllabus, which I shall be curious to look over. I never *enjoyed* any other lectures in my life, except your's, for Edinburgh completely sickened me of that method of learning. What a grand step it would be to break down the system of eternal classics, & nothing but classics.— I am perfectly certain, that the only thing at Cambridge which did my mind any good, were your lectures & still more your conversation; I believe I must except, also, getting up Paley's Evidences. It would, indeed, be a grand step to get a little more diversity in study for men of different minds. Talking of classics reminds me to ask you to do me a very *essential* favour: I find I have utterly forgotten my whole immense stock of classical knowledge which put me in the eminent position of 5th or 6th in the oi polloi.[5] Now I have to invent many names of families & genera for my work on Cirripedia, & I have not the **smallest** idea whether my names are correct. W^d. you let me send them to you, for your opinion from time to time? My paper c^d be returned with your fiat, so that you w^d have but little writing, or indeed hardly any.—

When you are walking on the shore, w^d you take the trouble to scrape me off (taking care to get the base) a few Barnacles, of as many different forms as you can see, & send them me in little strong box, damp with sea-weed. I am anxious to make out the distribution of the British species— And new species may turn up, for the group has been made out most superficially,—for instance under Balanus punctatus (which must be made a distinct genus) three or four varieties have been called distinct species; whereas one form, which has not been called even a variety, is not only a distinct genus, but a distinct sub-family.— Yesterday I found four or 5 named genera are all the closest species of one genus: this will give you a specimen of the utter confusion my poor dear Barnacles are in.—

I am in a very cock-a-hoop state about my anatomy of the Cirripedia, & think I have made out some very curious points: my Book will be published in two years by the Ray Soc. & will I trust do no discredit (see how vain I am!) to your old pupil & most attached friend

    C. Darwin

To Emma Darwin   [17 November 1848]

                  Park St.—

                  3 oclock

My own ever dear Mammy.—

Here I am & have had some tea & toast for luncheon & am feeling very well. My drive did me good & I did not feel exhausted till I got near here & now I am rested again & feel pretty nearly at my average.—⁶

My own dear wife, I cannot possibly say how beyound all value your sympathy & affection is to me.— I often fear I must wear you with my unwellnesses & complaints.

Your poor old Husband | C. D.

1849

To W. D. Fox   6 February [1849]

Down Farnborough Kent
Feb 6th.

My dear Fox

I was very glad to get your note. I have often been thinking of writing to you, but all the autumn & winter I have been much dispirited & inclined to do nothing but what I was forced to.

I saw two very nice notes of yours on the ocasion of my poor dear Father's death. The memory of such a Father is a treasure to one; & when last I saw him he was very comfortable & his expression which I have now in my mind's eye serene & cheerful.—

Thank you much for your information about the water cure: I cannot make up my mind; I dislike the thoughts of it much— I know I shall be very uncomfortable there, & such a job moving with 6 children. Can you tell me (& I sh^d be much obliged sometime for an answer) whether either your cases was dyspepsia, though D^r. Holland does not consider my case quite that, but nearer to suppressed gout. He says he never saw such a case, & will not take on him to recommend the water cure.— I must get Gully's Book.— . . .

Pray give our kind remembrances to Mrs Fox, & believe my dear old Friend | Yours most sincerely | C. Darwin

[From 10 March to 30 June, the Darwin family took lodgings in Malvern, Worcestershire, where CD undertook Dr John Manby Gully's watercure.]

To Susan Darwin   [19 March 1849]

[Malvern]
Monday

My dear Susan.

As you say you want my hydropathical diary, I will give it you —though tomorrow it is to change to a certain extent.— $\frac{1}{4}$ before 7. get up, & am scrubbed with rough towel in cold water for 2 or

105

3 minutes, which after the few first days, made & makes me very like a lobster— I have a washerman, a very nice person, & he scrubs behind, whilst I scrub in front.— drink a tumbler of water & get my clothes on as quick as possible & walk for 20 minutes— I c^d. walk further, but I find it tires me afterwards— I like all this very much.— At same time I put on a compress, which is a broad wet folded linen covered by mackintosh & which is "refreshed"—ie dipt in cold water every 2 hours & I wear it all day, except for about 2 hours after midday dinner; I don't perceive much effect from this of any kind.— After my walk, shave & wash & get my breakfast, which was to have been exclusively toast with meat or egg, but he has allowed me a little milk to sop the *stale* toast in. At no time must I take any sugar, butter, spices tea bacon or anything good.— At 12 oclock I put my feet for 10 minutes in cold water with a little mustard & they are violently rubbed by my man; the coldness makes my feet ache much, but upon the whole my feet are certainly less cold than formerly.— Walk for 20 minutes & dine at one.— He has relaxed a little about my dinner & says I may try plain pudding, if I am sure it lessens sickness.—

After dinner lie down & try to go to sleep for one hour.— At 5 olock feet in cold water—drink cold water & walk as before— Supper same as breakfast at 6 oclock.— I have had much sickness this week, but certainly I have felt much stronger & the sickness has depressed me much less.— Tomorrow I am to be packed at 6 oclock A.M for 1 & $\frac{1}{2}$ hour in Blanket, with hot bottle to my feet & then rubbed with cold dripping sheet; but I do not know anything about this.— I grieve to say that Dr Gully gives me homœopathic medicines three times a day, which I take obediently without an atom of faith.— I like Dr Gully much— he is certainly an able man: I have been struck with how many remarks he has made similar to those of my Father.—

He is very kind & attentive; but seems puzzled with my case— thinks my head or top of spinal chord cause of mischief— He has generously allowed me 6 pinches of snuff for all this week, which is my chief comfort except thinking all day of myself & prosing to Emma, who bless her old soul, thinks as much about me as I do even myself.— I am become perfectly indolent which I feel the oddest change of all to myself & this letter is the greatest mental effort done by me since coming here— My dearest sisters I wish I c^d. see you here.— I saw absolutely nothing of you at Down & never talked about my dear Father about whom it is now to me the sweetest pleasure to think, which I fear cannot be your case as yet.

My dears | Yours affectionly | C. D. . . .

To Syms Covington   30 March 1849

Down Farnborough, Kent, [Malvern]

March 30, 1849.

Dear Covington,—

It is now some years since I have heard from you, and I hope you will take the trouble to write to me to tell me how you and your family are going on. I should much like to hear that your worldly circumstances are in a good position, and that you are every way fortunate. I hope that your deafness has not increased. I will now tell you about myself. My poor dear father, whom you will remember at Shrewsbury, died in his 84th year on the 13th of November. My health lately has been very bad, and I thought all this winter that I should not recover. I am now not at home (though I have so dated this letter) but have come to Malvern for two months to try the cold water cure, and I have already received so much benefit that I really hope my health will be much renovated. I have finished my three geological volumes on the voyage of the old Beagle, and my journal, which you copied, has come out in a second edition, and has had a very large sale. I am now employed on a large volume, describing the anatomy and all the species of barnacles from all over the world. I do not know whether you live near the sea, but if so I should be very glad if you would collect me any that adhere (small and large) to the coast rocks or to shells or to corals thrown up by gales, and send them to me without cleaning out the animals, and taking care of the bases. You will remember that barnacles are conical little shells, with a sort of four-valved lid on the top. There are others with long flexible footstalk, fixed to floating objects, and sometimes cast on shore. I should be very glad of any specimens, but do not give yourself much trouble about them. If you do send me any, they had better be directed to the Geological Society, Somerset House, and a letter sent to inform me of them. I shall not publish my book for 18 months more.

I have now six children—three boys and three girls—and all, thank God, well and strong. I have not seen any of our old officers for a long time. Captain Fitz Roy has the command of a fine steamer frigate. Captain Sulivan has gone out to settle for a few years, and trade at the Falkland Islands, and taken his family with him. I know nothing of the others. You will remember Evans, my father's butler at Shrewsbury; he and his wife are both dead. I should like to hear what you think of the prospects of your country. How is Captain King? Should you see Mr. Philip King, please say that I desired to be most kindly remembered to him; I was grieved to hear some long time since that he was out of health. Has he any family? I often think how many pleasant walks I had with him. Speaking of walks, I fear my day is done, and I could never tire you again. I have not been able to walk a mile for some years, but now with the water-cure I am getting stronger again. With every hope that you are happy and prosperous, believe me, dear Covington, your sincere well-wisher, C. Darwin.

To Charles Lyell [14–28 June 1849]

The Lodge Malvern
Friday

My dear Lyell

We were uncommonly much obliged to Lady Lyell for her most agreeable letter which told us much which we were very glad & curious to hear. Emma has deputed me to write, for she, poor soul, is in her usual wretched state, which to none of our friends requires any further explanation.—

I have got your Book[1] & have read all first & small part of 2^d. Volume (reading is the hardest work allowed here) & greatly I have been interested by it— It makes me long to be a Yankey.— Emma desires me to say that she quite "gloated" over the truth of your remarks on religious progress; lying sick on the sofa it has been the only Book she has much enjoyed for a long time. I delight to think how you will disgust some of the Bigots & Educational Dons. . . .

We return home on 30th inst. I have not been quite so well the last week; but I had a few days before that of almost perfect health: the D^r. thinks he can quite cure me, but I must go on with all the processes for several more months & he urges me to keep perfectly idle for some time longer, which is a great bore, though it is wonderful how one gets accustomed to any thing: I have bought a horse & taken to ride.— If I go on very well I shall certainly be at Birmingham;[2] but otherwise not, for I am determined to try my best & get decent health again. . . .

I shall be astonished if your Book has not an immense sale, for almost everyone is interested about America, & all who are, must enjoy your Book

Yours most sincerely | C. Darwin

To Charles Lyell   [2 September 1849]

Down Farnborough Kent
Sunday

My dear Lyell

... We are going on as usual: Emma desires her kind love to Lady Lyell: she boldly means to come to Birmingham with me & very glad she is that Lady Lyell will be there: two of our children have had a tedious slow fever.— I go on with my aqueous processes & very steadily but slowly gain health & strength. Against all rules I dined at Chevening with L^d. Mahon, who did me the gr^t. honour of calling on me, & how he heard of me, I can't guess— I was charmed with Lady Mahon, & anyone might have been proud at the praises of agreeableness which came from her beautiful lips with respect to you.— I liked old L^d. Stanhope very much; though he abused geology & zoology heartily— "To suppose that the omnipotent God made a world, found it a failure, & broke it up & then made it again & again broke it up, as the geologists say, is all fiddle faddle".— Describing *species* of birds & shells &c is all "fiddle faddle". But yet I somehow liked him better than Ld Mahon.—

I am heartily glad we shall meet at Birmingham, as I trust we shall if my health will but keep up.— I work now every day at the Cirripedia for $2\frac{1}{2}$ hours & so get on a little but very slowly.— I sometimes after being a whole week employed & having described, perhaps only 2 species agree mentally with L^d. Stanhope that it is all fiddle-faddle: however the other day I got the curious case of a unisexual, instead of hermaphrodite, cirripede, in which the female had the common cirripedial character, & in two of the valves of her shell had two little pockets, in *each* of which she kept a little husband; I do not know of any other case where a female invariably has two husbands.— I have one still odder fact, common to several species, namely that though they are hermaphrodite, they have small additional or as I shall call them Complemental males:[3] one specimen itself hermaphrodite had no less than *seven* of these complemental males attached to it. Truly the schemes & wonders of nature are illimitable.— But I am running on as badly about my Cirripedia as about Geology: it makes me groan to think that probably, I shall never again have the exquisite pleasure of making out some new district,—of evoking geological light out of some troubled, dark region.— So I must make the best of my Cirripedia. ...

Yours most sincerely | C. Darwin

To J. D. Hooker   12 October 1849

Down Farnborough Kent
Oct 12. 49

My dear Hooker

I was heartily glad to get your last letter; but on my life your thanks for my very few & very dull letters quite scalded me— I have been very indolent & selfish in not having oftener written to you & kept my ears open for news which w^d. have interested you; but I have not forgotten you. Two days after receiving your letter, there was a short leading notice about you, in Gardener's Chron^e; in which it is said that you have discovered a noble crimson rose & 30 Rhododendrons: I most heartily congratulate you on these discoveries, which will interest the Public; & I have no doubt that you will have made plenty of most interesting botanical observations.— This last letter shall be put with all your others which are now safe together.—

I am very glad that you have got minute details about the terraces in the valleys; your description sounds curiously like the terraces in the Cordillera of Chile; these latter, however, are single in each valley; but you will hereafter see a description of these terraces in my Geolog. of S. America. ...

You ask about my Cold Water Cure; I am going on very well & am certainly a little better every month; my nights mend much slower than my days.— I have built a douche & am to go on through all the winter, frost or no frost— My treatment now is lamp 5 times per week & shallow bath for 5 minutes afterwards; douche daily for 5 minutes & dripping sheet daily. The treatment is wonderfully tonic, & I have had more better consecutive days this month, than on any previous ones.— The vomiting I consider absolutely cured. I am allowed to work now $2\frac{1}{2}$ hours daily, & I find it as much as I can do; for the cold-water cure, together with 3 short walks is curiously exhausting; & I am actually *forced* always to go to bed at 8 oclock completely tired.— I steadily gain in weight & eat immensely & am never oppressed with my food. I have lost the involuntary twitching of the muscles & all the fainting feelings &c black spots before eyes &c &c Dr Gully thinks he shall quite cure me in 6 or 9 months more.—

The greatest bore, which I find in the Water Cure, is the having been compelled to give up all reading, except the newspapers; for my daily $2\frac{1}{2}$ hours at the Barnacles is fully as much as I can do of anything which occupies the mind: I am consequently terribly behind in all Scientific books.—

I have of late been at work at mere species describing, which is much more difficult than I expected & has much the same sort of interest as a puzzle has; but I confess I often feel wearied with the work & cannot help sometimes asking myself what is the good of spending a week or fortnight in ascertaining that certain just perceptible differences blend together & constitute varieties & not species. As long as I am on anatomy I never feel myself in that disgusting, horrid cui bono inquiring humour. What miserable work, again, it is searching for priority of names; I have just finished two species which possess seven generic & 24 specific names! My chief comfort is, that the work must be sometime done, & I may as well do it, as anyone else.— I have given up my agitation against mihi & nobis; my paper is too long to send to you, so you must see it, if you care to do so, on your return. By the way, you say in your letter that you care more for my species work than for the Barnacles; now this is too bad of you, for I declare your decided approval of my plain Barnacle work over theoretic species work, had very great influence in deciding me to go on with former & defer my species-paper. ...

My wife desires to be most kindly remembered to you: she will be confined of our no^r 7 in January; is this not a dreadful number? all the six are well & lively. ...

Farewell my dear Hooker with every good wish. | Yours affectionately | C. Darwin ...

[Earlier in 1849, CD had initiated a correspondence with James Dwight Dana to enlist his aid with his work on barnacles.]

To James Dwight Dana   5 December [1849]

> Down Farnborough Kent
> Dec. 5th

My dear Sir

I have not for some years been so much pleased, as I have just been by reading your most able discussion on coral-reefs.— I thank you

most sincerely for the very honourable mention you make of me.—
This day I heard that the Atlas[4] has arrived & this completes your
munificent present to me.— I have not yet come to the chapter on
subsidence, & in that I fancy we shall disagree, but in the descriptive
part, our agreement has been eminently satisfactory to me, & far
more than I ever ventured to anticipate.— I consider that *now* the
subsidence theory is established.—

I have read about half through the descriptive part of the volcanic
Geology (last night I ascended the peaks of Tahiti with you, & what
I saw in my short excursion was most vividly brought before me
by your descriptions) & have been most deeply interested by it: your
observations on the Sandwich craters strike me as the most important
& original of any that I have read for a long time. Now that I have
read you, I believe I saw at the Galapagos, at a distance, instances
of those most curious fissures of eruption. There are many points of
resemblance between the Galapagos & Sandwich is^{ds} (even to the
shape of the mound-like hills) viz in the liquidity of the lavas, absence
of scoriæ & [presence of] tuff-craters.— Many of your scattered
remarks on denudation have particularly interested me; but I see
that you attribute less to sea & more to running water than I have
been accustomed to do.— After your remarks in your last very kind
letter, I could not help skipping on to the Australian valleys, on which
your remarks strike me as exceedingly ingenious & novel, but they
have not converted me: I cannot conceive how the great *lateral bays*
could have been scooped out, & their sides rendered precipitous by
running water. I shall go on & read every word of your excellent
volume. . . .

When I meet a very goodnatured man, I have that degree of bad-
ness of disposition in me, that I always endeavour to take advantage
of him: therefore I am going to mention some desiderata, which if
you can supply I shall be very grateful, *but if not no answer will be
required*. I want much a specimen of Coronula denticulata of Say on
the Kings Crab of U. States.— I *especially* want any species of the
genus Scalpellum (of course I w^{d}· return any specimens only lent me,
only I require to open one specimen of each kind). How far south in
Antarctic sea did you meet any Anatifas; I have Sir J. Ross' collection but there are no precise localities: the common Antartic Anatifa
I have called australis; it differs from all northern forms.—

Do you know any Crustacean which bores in calcareous rocks or
shells? Do you know any crust, whose oviducts open at or near the

antennæ? Did you discover where the ovaria are situated in Phyllosoma? Lastly can you tell me whether any list has been published of the plants found on **Elevated** coral islands; or could you procure me such a list.— Now can you forgive me asking you all these questions; please observe, that I beg you not to answer, without you can inform me on these points (which I well know is not likely) or help me with respect to above specimens. . . .

With my most sincere thanks for the honour you have done me & the gratification you have afforded me | Pray believe me | Yours very sincerely | C. Darwin . . .

1850

[Hooker and his travelling companion, Andrew Campbell, political agent to Sikkim, had been taken captive in November 1849 by a Sikkim band under instructions from the Sikkim Dewan. Both were released in December after a threat of force by the British Government.]

To J. D. Hooker   3 February [1850]

Down Farnborough Kent
Feb. 3^{d}.

My dear Hooker

I hope that there cannot be a shadow of a doubt, that long before this letter gets to India, you will be a free man.— I was quite astounded a short time ago, when indolently skimming through the Paper, to see your imprisonment announced. I was at first anxious enough about your safety, so I wrote to Henslow, & this, through the very great kindness of your Father, procured me a note from him, giving me details of all that he knew, & which, I trust, shows that your case is not bad. Indeed if you are enabled to go on collecting it may even be a good thing. In another way, I hope it may be a good thing, for perhaps Sir William & Lady Hooker will insist on your coming home;— surely you must have reaped a noble & sufficient Botanical harvest.— For myself I have in truth no news; I have never been so much cut off from all scientific friends, for I have found that interrupting the water cure does not answer. My health has of late kept stationary, & I begin to fear I shall not derive much more benefit from W. Cure; though the amount has been more than I at first even dared to hope for.— Sharp work my Baths have been for 5 minutes with water under 40°.— I am on the Council of Royal Soc. & am ashamed to say that I have not attended once.—

I have now for a long time been at work at the fossil cirripedes, which take up more time even than the recent;—confound & exterminate the whole tribe; I can see no end to my work.—

My wife desires her kindest remembrances to you; she has lately produced our fourth Boy & seventh child![1]— a precious lot of young beggars we are rearing.— I was very bold & administered myself, before the Doctor came, Chloroform to my wife with admirable success.— ...

Farewell, forgive this dull letter, & accept all good wishes of all kinds from your most sincere friend | C. Darwin ...

To W. D. Fox   4 September [1850]

Down Farnborough Kent
Sept. 4th.—

My dear Fox

I was much pleased to get your very agreeable letter with all its curious facts on the female sex & their hereditariness. ... I wonder whether the queries addressed to about the specific distinctions of the races of man are a reflexion from Agassiz's Lectures in the U.S. in which he has been maintaining the doctrine of several species, —much, I daresay, to the comfort of the slave-holding Southerns.— Your aphorism that "any remedy will cure any malady" contains, I do believe, profound truth,—whether applicable or not to the wondrous Water Cure I am not very sure.— The Water-Cure, however, keeps in high favour, & I go regularly on with douching &c &c: I am much in the same state as I have been for the last nine months, & not quite so brilliantly well as I was in the dead of last winter. To be as I am, though I never have my stomach right for 24 hours, is, compared to my state two years ago, of inestimable value.

My wife & all my children are well; & they, the children, are now seven in number; to what I am to bring up my four Boys, even already sorely perplexes me. My eldest boy is showing the hereditary principle, by a passion for collecting Lepidoptera. We are at present very full of the subject of schools; I cannot endure to think of sending my Boys to waste 7 or 8 years in making miserable Latin verses, & we have heard some good of Bruce Castle School, near Tottenham which is partly on the Fellenberg System, & is kept by a Brother of Rowland Hill of the Post-office, so that on Friday we are going to inspect it & the Boys. I feel that it is an awful experiment to depart from the usual course, however bad that course may be.— Have you, who have something of an omniscient tendency in you, ever heard anything of this school?—

You speak about Homœopathy; which is a subject which makes me more wrath, even than does Clair-voyance: clairvoyance so transcends belief, that one's ordinary faculties are put out of question, but in Homœopathy common sense & common observation come into play, & both these must go to the Dogs, if the infinetesimal doses have any effect whatever. How true is a remark I saw the other day by Quetelet, in respect to evidence of curative processes, viz that no one knows in disease what is the simple result of nothing being done, as a standard with which to compare Homœopathy & all other such things. It is a sad flaw, I cannot but think in my beloved Dr Gully, that he believes in everything— when his daughter was very ill, he had a clair-voyant girl to report on internal changes, a mesmerist to put her to sleep—an homœopathist, viz D^r. Chapman; & himself as Hydropathist! & the girl recovered.—

My dear Fox, I do hope we shall sometime see you here again. Your affectionate friend | C. Darwin . . .

To W. D. Fox   10 October [1850]

> Down Farnborough Kent
> Oct. 10th

My dear Fox

I am very much obliged for your juicy, as my poor dear Father used to call an interesting, letter.— We were very glad to get the sentence about Bruce Castle school, for we are still in an awesome state of indecision between Rugby & it. I knew you were just the man to apply to to get information upon any out of the way subject.— We have taken much pains in making enquiries, & upon the whole the balance is decidedly favourable; yet there is so much novelty in the system that we cannot help being much afraid at trying an experiment on so important a subject. At Bruce castle, they do not begin Latin, till a Boy can read, write, spell, & count *well*: they have no punishments except stopping premiums on good behaviour. I do not see how we are ever to come to a decision; but we must soon.— Willy is 11 this coming Christmas, & backward for his age; though sensible & observant. I rather think we shall send him to Bruce C. School.— Your own system of Education sounds **capital**, & why you sh^d. think I sh^d. laugh at it, I cannot conceive: I believe a good deal of diversity an immense advantage. It is one good point at Bruce Castle: that no one subject exceeds an hour & if a Boy can do it quicker, he may go out before the hour is over.—

You say you are teaching riding: we have been teaching Willy & we began without stirrrups, & in consequence Willy got two *severe* falls, one almost serious; so we are thinking of giving him stirrups;

more especially as I am assured, that a Boy who rides well without
stirrups has almost to begin again when he takes to stirrups: Can you
give me any wisdom on this head; pray do if you can?— ...

Yours affectionately | C. Darwin

Do you intend to educate your Boys altogether at home?— The
first-rate tutor at whom Willy now is, teaches nothing on earth but
the Latin Grammar, & his charge is 150£ per annum! Bruce Castle
is cheap *with extras* about 80£. ...

[The final decision was to send William to Rugby School. He entered the school in
early February 1852. In a letter to W. D. Fox, dated 7 March [1852], CD confessed
that he had not had the courage to break away from 'the old stereotyped stupid
classical education'.]

To Syms Covington   23 November 1850

> Down Farnborough, Kent,
> November 23, 1850.

Dear Covington,—

I received your letter of the 12th of March on the 25th of August,
but the box of which you advised me arrived here only yesterday.
The captain who brought it made no charge, and it arrived quite
safely. I thank you very sincerely for the great trouble you must have
taken in collecting so many specimens. I have received a vast number
of collections from different places, but never one so rich from one
locality. One of the kinds is most curious. It is a new species of a genus
of which only one specimen is known to exist in the world, and it is
in the British Museum. I see that you are one of those very rare few
who will work as hard for a friend when several thousand miles apart
as when close at hand. There are at least seven different kinds in the
box. The collection must have caused you much time and labour,
and I again thank you very sincerely for so kindly obliging me. I have
been amused by looking over two old papers you used in packing
up, and in seeing the names of Captain Wickham, Mr. Macleay, and
others mentioned. I am always much interested by your letters, and
take a very sincere pleasure in hearing how you get on. You have an
immense, incalculable advantage in living in a country in which your
children are sure to get on if industrious. I assure you that, though
I am a rich man, when I think of the future I very often ardently
wish I was settled in one of our Colonies, for I have now four sons
(seven children in all, and more coming), and what on earth to bring
them up to I do not know. A young man may here slave for years

in any profession and not make a penny. Many people think that Californian gold will half ruin all those who live on the interest of accumulated gold or capital, and if that does happen I will certainly emigrate. Whenever you write again tell me how far you think a gentleman with capital would get on in New South Wales. I have heard that gentlemen generally get on badly. I am sorry to say that my health keeps indifferent, and I have given up all hopes of ever being a strong man again. I am forced to live the life of a hermit, but natural history fills up my time, and I am happy in having an excellent wife and children. Any particulars you choose to tell me about yourself always interest me much. What interest can you get for money in a safe investment? How dear is food; I suppose nearly as dear as in England? How much land have you? I was pleased to see the other day that you have a railway commenced, and before they have one in any part of Italy or Turkey. The English certainly are a noble race, and a grand thing it is that we have got securely hold of Australia and New Zealand.

Once again accept my thanks for your valuable collection of barnacles, and believe me, dear Covington, your sincere friend, C. Darwin.

1851

[Darwin's ailing eldest daughter Anne Elizabeth was undergoing treatment at Malvern. When her condition deteriorated, Darwin was summoned to her bedside. Emma, in the late stages of pregnancy, remained at Down.]

To Emma Darwin   [17 April 1851]

[Malvern]
4 oclock.

My dearest Emma.

I am assured that Annie is several degrees better: I have in vain tryed to see D.ʳ Gully as yet. She looks very ill: her face lighted up & she certainly knew me.— She has not had wine, but several spoon-fulls of broth, & ordinary physic of camphor & ammonia— Dʳ Gully is most confident there is strong hope.— Thank God she does not suffer at all—half dozes all day long. I will write again if I can anyhow see Dʳ Gully before seven oclock. My own dearest support yourself— on no account *for the sake of ⟨ou⟩r other children*; **I implore you**, do not think of coming here.—

Yours my dearest | C. Darwin

I am assured there is great hope.— Yesterday she was a little better, & today again a little better.—

To Emma Darwin   [20 April 1851]

[Malvern]
Sunday.

My dear Emma

I had not time to send a second later letter yesterday. I do not know, but think it is best for you to know how every hour passes. It is a relief to me to tell you: for whilst writing to you, I can cry; tranquilly. I forget whether I told you that she vomited yesterday evening & *slightly* a second time. A second injection produced no sort of effect & did not relieve, but seems unimportant We then had to get Surgeon to draw her water off: this was done well & did not hurt her, but she struggled with surprising strength against being

119

uncovered &c. soon it evidently relieved her. All night she has slept tranquilly except for about 10 minutes, when she wandered in slightly excited manner. Dr G. came at 11°. 30′ & again said not worse. She has, however, taken less gruel this night & is fearfully prostrated. Yet when Brodie sponged her face, she asked to have her hands done and then thanked Brodie. & put her arms round her neck, my poor child & kissed her—

She vomited a mouthful this morning. It is certain she suffers very little—dosing nearly all the time: occasionally she says she is very weak. I expect D^r. G. immediately. Last night D^r. G. said, "you must not trust me, for I can give no reason for my intuition, but yet I think she will recover" Fanny H. sat up till 2 oclock God bless her. she is most sympathetic yet encouraging. Poor dear devoted Miss Thorley thus had one entire nights rest.—

8 oclock. A.M. Dr G. has been & again he says *positively* no symptom is worse, but none better: he cares less about food than I expected: if she can weather the fortnight, he has some hopes.— Your two heart-moving notes have come. My dear dear wife.— I do not sit all the while, with her, but am constantly up & down: I *cannot* sit still.—

10 oclock. I grieve to say she has vomited rather much again: but Mr Coates has been & drawn off again *much water* & this he says is a very good symptom. Last night he seemed astonished at her "fearful illness" & he made me very low; so this morning I asked nothing & he then felt her pulse of his own accord & at once said, "I declare I almost think she will recover". Oh my dear was not this joyous to hear.— He then went on to say (& I believe him from what my Father has said) that Fever at the same period is generally either fatal to many or though appearing very bad does not kill one: & now he himself has had 6 or 7 *most severe* cases in the low country beneath Malvern & not one has died.—

She has her senses remarkably today which is very good as showing head not affected: she called Papa when I was out of room unfortunately & then added "is he out?" This & her speeches to Brodie show more clearness of mind than I have seen, & she knew what M^r. Coates was going to do.— Several of M^r. Coates fever patients have had their bladders paralysed the whole time. Oh I do wish for Tuesday the fortnight to be over.— But I must not hope too much.— These alternations of no hope & hope sicken one's soul: I cannot help getting so sanguine everynow & then to be disappointed.

12 oclock. Again she has vomited & complains of fatigue rather more. She is very sensible; I was moving her, when she said "Dont do that please" & when I stopped "thank you".— 2 oclock, again she has vomited but again D^r. G. who has just been here says her pulse is rather better, certainly not worse.— We have put mustard poultice on stomach, & that has smarted her a good deal,—which shows more sensibility than I expected.— 3 oclock; she is a little chilly & we have given her a little Brandy—& hope she is asleep & I trust will warm.— I never saw anything so pathetic as her patience & thankfulness; when I gave her some water, she said "I quite thank you".— Poor dear darling child. The D^r. will come at 7 again.—

4°. 30$'$ The chilliness has pretty well gone off & no more sickness, refreshing sleep.

I will write again, if I have time | Yours | C. D

To Emma Darwin   [23 April 1851]

[Malvern]
Wednesday

My dear dearest Emma

I pray God Fanny's note may have prepared you. She went to her final sleep most tranquilly, most sweetly at 12 oclock today. Our poor dear dear child has had a very short life but I trust happy, & God only knows what miseries might have been in store for her. She expired without a sigh. How desolate it makes one to think of her frank cordial manners. I am so thankful for the daguerreotype. I cannot remember ever seeing the dear child naughty. God bless her. We must be more & more to each other my dear wife— Do what you can to bear up & think how invariably kind & tender you have been to her.— I am in bed not very well with my stomach. When I shall return I cannot yet say. My own poor dear dear wife.

C. Darwin

[CD, anxious about Emma's condition, left the funeral arrangements to relatives and returned to Down. Horace Darwin was born on 13 May 1851. CD resumed his work on barnacles.]

To William Erasmus Darwin   3 October [1851]

Down
Friday. Octr. 3^d.

My dear old Willy

I have been intending for some time to write to you, but had not much to say, & that is pretty nearly the case at present.— Mamma was to have started for Shrewsbury & Barlaston on Saturday, but the Baby is not quite well, so she has put it off for a week. Georgy is terribly disappointed, for he now as much likes going from home, as he formerly did not care about it. All day long Georgy is drawing ships or soldiers, more especially drummers, whom he will talk about as long as anyone will listen to him.— ...

I hope that you will think *sometimes* about what we talked about one of the last mornings.— You will surely find that the greatest pleasure in life is in being beloved; & this depends almost more on pleasant manners, than on being kind with grave & gruff manners. You are almost always kind & only want the more easily acquired external appearance. Depend upon it, that the only way to acquire pleasant manners is to try to please *everybody* you come near, your school-fellows, servants & everyone. Do, my own dear Boy, sometimes think over this, for you have plenty of sense & observation.

Love from Mamma. | Your's affectionately. | C. Darwin

1852–1854

To W. E. Darwin   24 [February 1852]

Down.

Tuesday 24th.—

My dear old Willy

I have not for a very long time been more pleased than I was this morning at receiving your letter with the *excellent* news at your having got so good a place [at Rugby]. We are both rejoiced at it, & give you our hearty congratulations. It is in every respect a very good thing, for you will be amongst an older set of Boys. Your letter was a very good one, & told us all that we liked to hear: it was well expressed & you must have taken some pains to write it. We are so very glad to hear that you are happy & comfortable; long may you keep so my dear Boy.— What a tremendous, awful, stunning, dreadful, terrible, bothering steeple-chase, you have run: I am astonished at your getting in the the 5th.. When next you write, explain how it came that you, a new Boy, & Erny, an old Boy, came to run together? What boys run, all those in your house? or in your Form? . . .

Next Sunday when you write here, tell us who your master is, & what Books you are in; & give us a History of the last Friday. The more you can write, the better we shall be pleased. All the servants enquire about you; and so they did at Aunt Sarah's. I was there the other day & saw the pretty little grey Hens. We are doing nothing particular; one day is like another: I go my morning walk & often think of you, & Georgy draws every day many Horse-guards; and Lizzie shivers & makes as many extraordinary grimaces as ever, & Lenny is as fat as ever. We shall probably come & see you during the first week in April. . . .

Farewell my dear Willy; may you go on as well as you have begun. All here send their best loves to you. | Your affectionate Father | C. Darwin . . .

I was saying before Georgy that he did not much like reading, when he said "No, I hate reading, but I like money."— I suppose he thought this made up for his not liking reading.—

To W. D. Fox   7 March [1852]

Down Farnborough Kent
March. 7th.

My dear Fox.

It is indeed an age since we have had any communication, & very glad I was to receive your note. Our long silence occurred to me a few weeks since, & I had then thought of writing but was idle. I congratulate & condole with you on your *tenth* child; but please to observe when I have a 10th, send only condolences to me. We have now seven children, all well Thank God, as well as their mother; of these 7, five are Boys; & my Father used to say that it was certain, that a Boy gave as much trouble as three girls, so that bonâ fide we have 17 children.

It makes me sick whenever I think of professions; all seem hopelessly bad, & as yet I cannot see a ray of light.— I should very much like to talk over this (By the way my three Bug-bears are Californian & Australian Gold, beggaring me by making my money on mortgage worth nothing —The French coming by the Westerham & Sevenoaks roads, & therefore enclosing Down —and thirdly Professions for my Boys.) & I sh^d like to talk about Education, on which you ask me what we are doing. No one can more truly despise the old stereotyped stupid classical education than I do, but yet I have not had courage to break through the trammels. After many doubts we have just sent our eldest Boy to Rugby, where for his age he has been very well placed. By the way, I may mention for chance of hereafter your wishing for such a thing for any friends, that M^r. Wharton Vicar of Mitcham, appear to us a really excellent preparatory tutor or *small* school keeper.— I honour, admire & envy you for educating your Boys at home. What on earth shall you do with your Boys? . . .

Very many thanks for your most kind & large invitation to Delamere; but I fear we can hardly compass it. I dread going anywhere, on account of my stomach so easily failing under any excitement. I rarely even now go to London; not that I am at all worse, perhaps rather better & lead a very comfortable life with my 3 hours of daily work, but it is the life of a hermit. My nights are *always* bad, & that stops my becoming vigorous.— You ask about water cure: I take at intervals of 2 or 3 month, 5 or 6 weeks of *moderately* severe treatment, & always with good effect. ...

How paramount the future is to the present, when one is surrounded by children. My dread is hereditary ill-health. Even death is better for them.

My dear Fox your sincere friend | C. Darwin. ...

P.S. Susan has lately been working in a way, which I think truly heroic about the scandalous violation of the act against children climbing chimneys. We have set up a little Society in Shrewsbury to prosecute those who break the Law. It is all Susan's doing. She has had very nice letters from L^d. Shaftesbury & the D. of Sutherland, but the brutal Shropshire Squires are as hard as stone to move. The act out of London seems most commonly violated. It makes one shudder to fancy one of one's own children at 7 years old being forced up a chimney—to say nothing of the consequent loathsome disease, & ulcerlated limbs, & utter moral degradation. If you think strongly on this subject, do make some enquiries— add to your many good works—this other one, & try to stir up the magistrates. There are several people making a stir in different parts of England on this subject. ...

[CD had sent Dana and Johannes Peter Müller copies of the first volume of *Living Cirripedia*.]

To J. D. Dana   8 May [1852]

> Down Farnborough Kent
> May 8th

My dear Sir

Your letter has given me much pleasure,—more than you would anticipate, & more perhaps than it ought to do,—though I put down part of what you say to the kindness of disposition, which I have observed in your memoirs & in your letters to me. I have had a short letter from Müller of Berlin, expressing interest in my Book, and now, with what you have said, I feel highly satisfied, & can go on with my work with a good heart: You will perhaps be surprised at all this; but I think everyone wants sympathy in their pursuits, & I live a very retired life in the country, & for months together see no one out of my own large family. ...

I have lately been reading the vols. for the last dozen years of Silliman's Journal, with great interest: What a curious account is that on the blind Fauna by Mr Silliman, of the caves.— I feel extreme interest on the subject, having for many years collected facts on variation, &c &c.— Would it be possible to procure one of the Rats for the British Museum? I should so like my friend M^r. Waterhouse to examine the teeth & see whether it is an old or new world form. If ever you could oblige the naturalists on this side of the water by getting so interesting a specimen, would you send it to me to give to Waterhouse; for (**privately between ourselves**) it would be of little use to real science, if once in the hands of M^r. [J. E.] Gray; —but very likely I am asking for an impossibility; the rats may be very rare. It is not stated whether the optic nerve was dissected out, which would be a curious point. . . .

Accept my thanks for your very kind letter, & believe me | Very sincerely your's Charles Darwin.—

To John Higgins[1]   19 June [1852]

Down Farnborough Kent
June 19th.

My dear Sir

I beg to acknowledge & thank you for the Balance of £183" ^{s}4.^{d}11 now placed to my account at M^{rss} Robarts & Co.—

I should be much obliged if you would be so good, *when you have leisure*, as to let me hear what you think about my future prospects in regard to Rent. Now that wheat is not quite so low as it was & considering the prices of other products, fifteen per cent seems to me a large reduction, bearing in mind that the Farm buildings are new & that no timber or game is preserved. If the land when purchased was let at a very high rent, of course my remarks are not applicable.— As far as I can hear, 15 per cent is an unusually large reduction. I should like to hear what reduction the great landowners, namely, L^d. Yarborough & Mr Christopher in your district, have actually made.—

Although I am on principle a free-trader, of course I am not willing to make a larger reduction than necessary to retain a good tenant; on the other hand I sh^d be very sorry to press hardly on a tenant. Whatever reduction is to be made, I sh^d. think, (subject to your better judgment) had better, at least before long, be permanent on account of charges on the rent.

Nevertheless I must yet hope that agricultural produce will rise, for I believe prices on the continent, quite irrespectively of protection or free-trade, are below the average.— Pray do not until quite convenient trouble yourself by answering this. I will forward your answer to Miss Darwin who is interested on this point.

Pray believe me | My dear Sir | Your's very faithfully | Charles Darwin

J. Higgins Es^q

To Thomas Henry Huxley   11 April [1853]

Down, Farnborough, Kent.

April 11,

My dear Sir

I heard you say that you were at work at the Ascidiæ.[2] I have some 12–15 specimens in Spirits; I hope in fairish condition. It is very likely that you may have more than you want, but should you like my specimens they are completely at your service. It will give me some trouble to get them out of several large bottles, but it would give me *real pleasure* should you wish to have and examine them, but please do not say you should like them for mere form-sake. The colours are noted in some instances.

I procured a compound Ascidian (Boltenia?) at the Falklands (now I believe preserved in spirits) like a strawberry on a long foot-stalk; in this there were ova in all states which seemed to pass as they became mature out of what I considered the ovarium, into two gut-formed bags in each individual; and here they could be traced passing into larva, first with a long tail, (having transverse septa) coiled round the head or body, and then free, and causing the larva to be locomotive. In the same compound individual all the eggs and larvæ were in the same state; and when most matured, the animals were so shrunk, that the whole seemed formed of the gut-formed bags with the larvæ. In another genus (now dried) from T. del Fuego, there were also tailed larvæ. My descriptions were only such as an ignorant school-boy might make. Doubtless you have Müller's "Uber die Larven . . .[3] Echinodermen, *Vierte* Abhand:, 1852, Müller sent me a copy which is really wasted on me, and would be at the service of *anyone* who would value it.

You spoke as if you had had an intention to review my Cirripedia: it is very indelicate in me to say so, but it would give me *great* pleasure to see my work reviewed by any one so capable as you of praising anything which might deserve praise, and criticising the errors which no doubt it contains. My chief reason for wishing it, is, otherwise I do not believe any foreigner will ever hear of its existence. It has been published a year, and no notice has been taken of it by any zoologist, except briefly by Dana. Upon my honour I never did such a thing before as suggest (not that I have exactly suggested this time) a review to any human being. But having done so, I may mention that in my own opinion, the Limulus-like larva in 1st stage;—the mouthless pupa;—especially the method of cement with its modifications;—the senses;—& homologies & sexual peculiarities,—are the most curious points,—but I daresay I *greatly* exaggerate their curiosity, for I have become a man of one idea,—cirripedes morning & night.—

I am **perfectly** aware that with every wish on your part it may *easily* happen, that you could not spare time for *old* work, you having so much valuable *new* work.

Forgive the length & egotistical character of this note, & believe me | very truly your's | Charles Darwin

[Hooker had just published the introductory essay to *Flora Novæ Zelandiæ*, the second part of his *Botany of the Antarctic voyage* (London, 1853–5).]

To J. D. Hooker   25 September [1853]

Down Bromley Kent
Sept. 25th

My dear Hooker

I have read your paper with *great* interest. It seems all very clear; & will form an admirable introduction to the N.Z. Flora, or to any Flora in the world. How few generalisers there are amongst systematists; I really suspect there is something absolutely opposed to each other & hostile in the two frames of mind required for systematising, & reasoning on large collections of facts.— Many of your arguments appear to me very well put: & as far as my experience goes, the candid way in which you discuss the subject is unique. The whole will be very useful to me, whenever I undertake my volume; though parts take the wind very completely out of my sails, for I have for some time determined to give the arguments on *both* sides, (as far as I could) instead of arguing on the mutability side alone.—

I shall like very much seeing the remainder. ... In my own cirripedial work (by the way, thank you for the dose of soft solder, it does one, (or at least me) a great deal of good), —in my own work, I have not felt conscious that disbelieving in the *permanence* of species has made much difference one way or the other; in some few cases (if publishing avowedly on doctrine of non-permanence) I sh^{d}. *not* have affixed names, & in some few cases sh^{d}. have affixed names to remarkable varieties. Certainly I have felt it humiliating, discussing & doubting & examining over & over again, when in my own mind, the only doubt has been, whether the form varied *to-day or yesterday* (to put a fine point on it, as Snagsby[4] would say). After describing a set of forms, as distinct species, tearing up my M.S., & making them one species; tearing that up & making them separate, & then making them one again (which has happened to me) I have gnashed my teeth, cursed species, & asked what sin I had committed to be so punished: But I must confess, that perhaps nearly the same thing w^{d}. have happened to me on any scheme of work.— ...

Farewell, good luck to your work,—whether you make the species hold up their heads or hang them down, as long as you don't quite annihilate them or make them quite permanent; it will be all nuts to me; so farewell yours most truly | C. Darwin

[In a letter to CD dated [4 November 1853], Hooker stated: 'The R.S. have voted you the Royal Medal for Natural Science— *All along of the Barnacles*!!!'.]

To J. D. Hooker   5 November [1853]

Down Bromley Kent
Nov. 5th—

My dear Hooker

Amongst my letters received this morning, I opened first one from Col. Sabine: the contents certainly surprised me very much, but, though the letter was a *very kind one*, somehow, I cared very little indeed for the announcement it contained. I then opened yours, & such is the effect of warmth, friendship & kindness from one that is loved, that the very same fact told as you told it, made me glow with pleasure till my very heart throbbed. Believe me I shall not soon forget the pleasure of your letter. Such hearty affectionate sympathy is worth more than all the medals that ever were or will be coined. Again my dear Hooker, I thank you.—

I hope Lindley will never hear that he was a competitor against me; for really it is almost *ridiculous* (of course you would never repeat that I said this, for it would be thought by others, though not, I believe, by you, to be affectation) his not having the medal long before me; I must feel *sure*, that you did quite right to propose him; & what a good dear kind fellow you are, nevertheless, to rejoice in this honour being bestowed on me.

What *pleasure* I have felt on the occasion, I owe almost entirely to you.

Farewell my dear Hooker | yours affectionately | C. Darwin ...

To J. D. Hooker   27 [June 1854]

Down Farnborough Kent

27th.

My dear Hooker

I send you very sincere congratulations on your affair being over, in which my wife very truly joins.— You seem to have taken it very philosophically. In my opinion these affairs, like gales of wind, get less & less endurable.[5]

Did you administer the Chloroform? When I did, I was perfectly convinced that the Chloroform was very composing to oneself as well as to the patient. ...

With respect to "highness" & "lowness", my ideas are only *eclectic* & *not very clear*. It appears to me that an unavoidable wish to compare all animals with men, as supreme, causes some confusion; & I think that nothing besides some such vague comparison is intended, or perhaps is even possible, when the question is whether two kingdoms such as the articulata or mollusca are the highest. Within the same kingdom, I am inclined to think that "highest" usually means that form, which has undergone most "morphological differentation" from the common embryo or archetype of the class; but then every now & then one is bothered (as Milne Edwards has remarked) by "retrograde development", ie the mature animal having fewer & less important organs than its own embryo. The specialisation of parts to different functions, or "the division of physiological labour" of Milne Edwards exactly agrees (& to my mind is the best definition, when it can be applied) with what you state is your idea in regard to plants. I do not think zoologists agree in any definite ideas on this subject; & my ideas are not clearer than those of my Brethren.

Ever yours, C. Darwin ...

To J. D. Hooker   7 July [1854]

Down.

July 7th

My dear Hooker

I have had the House full of visitors, & when I talk I can do absolutely nothing else; & since then I have been poorly enough, otherwise I sh^d. have answered your letter long before this, for I enjoy extremely discussing such points, as those in your last note. But what a villain you are to heap gratuitous insults on my *elastic* theory; you might as well call the virtue of a lady *elastic*, as the virtue of a theory accomodating in its favours. Whatever you may say, I feel that my theory does give me some advantages in discussing these points:—

But to business, I keep my notes in such a way viz in bulk, that I cannot possibly lay my hand on any reference; nor as far as vegetable kingdom is concerneed do I distinctly remember having read any discussion on general highness or lowness, excepting Schleiden[6] (I fancy) on Compositæ being highest. Ad. de Jussieu in Arch. du Museum Tom. 8, discusses the value of characters of degraded flowers in the Malpighiaceæ, but I doubt whether this at all concerns you. Mirbel somewhere has discussed some such question.—

Plants lie under an enormous disadvantage in respect to such discussions in not passing through larval stages. I do not know whether you can distinguish a plant *low* from non development from one low from degradation, which theoretically, at least, are very distinct. I must agree with Forbes that *a* mollusc may be higher than one articulate animal & lower than another; if one was asked which was highest as a whole the Molluscan or Articulat Kingdom, I sh^d. look to & compare the highest in each, & not compare their archetypes (supposing them to be known, which they are not).—

But there are, in my opinion, more difficult cases, than any we have alluded to, viz that of Fish,—but my ideas are not clear enough & I do not suppose you w^d. care to hear what I obscurely think on this subject.— As far as my *elastic* theory goes all I care about is that very ancient organisms, (*when* different from existing,) sh^d tend to resemble the larval or embryological stages of the existing.— ...

This note is even feebler than my last, for I feel deadly sick, & decidedly an animal of low development.— I hope all goes on well at Hitcham.

Adios | C. Darwin ...

To T. H. Huxley   2 September [1854]

Down Farnborough Kent
Sept. 2^d.

My dear Sir

My second volume on the everlasting Barnacles is at last published, & I will do myself the pleasure of sending you a copy to

Jermyn St next Thursday, as I have to send another book there to M^r. Baily.— ...

I have just been reading your Review of the Vestiges, & the way you handle a great Professor is really exquisite & inimitable.[7] I have been extremely interested on other parts & to my mind it is *incomparably* the best review I have read on the Vestiges; but I cannot think but that you are rather hard on the poor author. I must think that such a book, if it does no other good, spreads the taste for natural science.—

But I am perhaps no fair judge for I am almost as unorthodox about species as the Vestiges itself, though I hope not *quite* so unphilosophical. How capitally you analyse his notion about law. I do not know when I have read a review which interested me so much. By Heavens how the blood must have gushed into the capillaries when a certain great man (whom with all his faults I cannot help liking) read it.—

I am rather sorry you do not think more of Agassizs embryological stages, for though I saw how excessively weak the evidence was, I was led to hope in its truth. I had no intention of prosing in this manner when I begun.

Pray believe me yours sincerely | C. Darwin

To Walter Baldock Durrant Mantell   17 November 1854

Down Bromley Kent
Nov. 17th. 1854

Dear Sir

I hope you will excuse the great liberty I take in addressing you, but I trust my long acquaintance with your honoured Father may serve me as an introduction. When on board H.M.S. Beagle I particularly attended to glacial deposits, & I am now very much interested on this subject, & most earnestly want to know whether any distinct phenomena of this kind have been observed in New Zealand.— When I was in Bay of Islands I saw several large boulders of greenstone, but as I did not know the surrounding country, & as they were in valleys & not on isolated hillocks, I was not able to tell whether they were true erratic boulders, or merely blocks washed down by chance floods &c &c from greater heights. Now I sh^d. esteem it a great favour if you would inform me on your own authority, (or on that of any other *competent* observer, if such there be in N. Zealand) whether there are any *great* blocks of rock, especially if *angular*, which have certainly been transported from a *long* distance, or which must have crossed valleys or arms of the sea in their course. And lastly whether any one has observed moraines or the marks of glaciers having formerly descended to a lower level on the New Zealand Mountains. The Southern islands w^d. of course be the most favourable for the chance of the discovery of ancient erratic & glacial action, but I fear that has been rarely visited by instructed persons.

Hoping that you will forgive my asking you to take the trouble to inform me; I remain, Dear Sir | Your's faithfully & obliged | Charles Darwin

I may add that on stating how anxious I was to obtain the foregoing information, I was encouraged by Sir Charles & Lady Lyell to take the liberty of addressing you.—

To J. D. Hooker   11 [December 1854]

Down Farnborough Kent
Monday 11

My dear Hooker

... With respect to splitting Australia we are in a "Muddle"; I do not think I quite understood you & you me; I am pretty sure I do not quite understand or remember what I wrote myself; & I doubt whether you quite understand or remember what you wrote; for in first letter you say "Under this view disruption produces similarity of Botanical features": in your second letter, you say that "three-fourths would be killed, & that a greater proportion of those species common to both (islands) would be killed, than of those peculiar to each"; but this w^d. produce *dis*similarity.—

Now for a short ride on my chief (at present) Hobby-Horse, viz aberrant genera: what you say under your remarks on Lepidodendron seems just the case, viz that I want to give some sort of evidence of what we both believe in, viz how groups come to be anomalous or aberrant. And I think some sort of proof is required; for I do not believe very many naturalists would at all admit our view. Thank you for caution on large anomalous genera first catching attention. I do not quite agree with your "grave objection to the whole process" which is "that if you multiply the anomalous species by 100, & divide the normal by the same, you will then reverse the names"...[8] — For, to take an example, ornithoryhnchus & Echidna would not be less aberrant if each had a dozen (I do not say 100, because we have no such cases in animal kingdom) species instead of one. What would really make these 2 genera less anomalous, would be the creation of *many genera* & sub-families round & radiating from them on all sides. Thus if Australia were destroyed Didelphys in S. America w^d be wonderfully anomalous (this is your case with Proteaceæ—), whereas now there are so many genera & little sub-families of Marsupiata, that the group cannot be called aberrant or anomalous. Sagitta (& the Earwig) is one of the most anomalous animals in world, & not a bit the less because there are a dozen species.— Now my point (which I think is a slightly new point of view) is, if it is extinction which has made the genus anomalous, *as a general rule*, the same causes of extinction would allow the existence of only a few species in such genera. . . .

I shall much like to hear whether this strikes you as sound; I feel all the time on the borders of a circle of truism.— Of course I could not think of such a request, but you might *possibly*, if Bentham does not think the whole subject rubbish, ask him sometime to pick out the dozen most anomalous genera in the Leguminosæ, or any *great* order of which there is a monograph, by which I could calculate the ordinary percentage of species to genera. I am the more anxious as the more I enquire, the fewer are the cases in which it can be done: it cannot be done in Birds or I fear Mammifers. I doubt much whether in any other class of Insects.— . . .

Farewell you most goodnatured of men. C. Darwin.—

I have just been testing practically what disuse does in reducing parts; I have made skeletons of wild & tame Duck (oh the smell of well-boiled, high Duck!!) & I find the tame-duck wing, ought according to scale of wild prototype to have its two wings 360 grams in weight, but it has it only 317 or 43 grams too little or $\frac{1}{7}$th of own two wings too little in weight: this seems rather interesting to me. . . .

1855

To J. D. Hooker   7 March [1855]

Down Farnborough Kent
March 7th.

My dear Hooker

... I have just finished working well at Wollaston's Insecta Mad: it is an *admirable* work. There is a very curious point in the astounding proportion of Coleoptera that are apterous; & I think I have grasped the reason, viz that powers of flight w^d be injurious to insects inhabiting a confined locality & expose them to be blown to the sea; to test this, I find that the insects inhabiting the Dezerta Grande, a quite small islet, would be still more exposed to this danger, & here the proportion of apterous insects is even considerably greater than on Madeira proper.—

Wollaston speaks of Madeira & the other archipelagoes as being "sure & certain witnesses of Forbes old continent," & of course the Entomological world implicitly follows this view. But to my eyes it w^d be difficult to imagine facts more opposed to such a view. It is really disgusting & humiliating to see directly opposite conclusions drawn from the same facts.— I have had some correspondence with W. on this & other subjects, & I find he coolly assumes (1) that formerly insects possessed greater migratory powers than now (2) that the old land was *specially* rich in centres of creation (3) that the uniting land was destroyed before the special creations had time to diffuse, & (4) that the land was broken down before certain families & genera had time to reach from Europe or Africa the points of land in question.— Are not these a jolly lot of assumptions? & yet I shall see for the next dozen or score of years Wollaston quoted as proving the former existence of poor Forbes' Atlantis.—

I hope I have not wearied you, but I thought you w^d. like to hear about this Book, which strikes me as *excellent* in its facts; & the Author a most nice & modest man.—

Most truly your's | C. Darwin

135

To W. D. Fox   27 March [1855]

Down Farnborough Kent
March 27th.

My dear Fox

... I forget whether I ever told you what the objects of my present work is,— it is to view all facts that I can master (eheu, eheu, how ignorant I find I am) in Nat. History, (as on geograph. distribution, palæontology, classification Hybridism, domestic animals & plants &c &c &c) to see how far they favour or are opposed to the notion that wild species are mutable or immutable: I mean with my utmost power to give all arguments & facts on both sides. I have a *number* of people helping me in every way, & giving me most valuable assistance; but I often doubt whether the subject will not quite overpower me.— ...

Farewell my dear Fox. ...

Your affectionate friend | C. Darwin ...

To J. D. Hooker   7 April [1855]

Down
Ap. 7th.

My dear Hooker

I wrote this morning to thank for the Rhododendrums.—

I have begun my seed-salting experiments, & I sh^d. be extremely much obliged if you would tell me what kinds you would expect to be *most easily killed* by sea-water besides the Cruciferæ, which I had thought w^d. be so, & which you confirmed; I had meant to have asked, but quite forgot, when I last saw you.—

If you can mention any that are easily procured, as Agricultural or Garden or flower seeds,—please enumerate **Just a few**.— Secondly will you tell me, at a *guess*, how long an immersion in sea-water you sh^d. *imagine* w^d. kill the more susceptible seeds? Should you expect a week's fair immersion w^d. destroy any of them? ...

Will you be so kind as to send me a brief note in answer, as I may thus be sooner put out of my pain, & end my experiments, which I daresay you think as foolish, as my splendid idea, that the Coal-plants lived in salt-water like mangroves which made you so savage

Adios | C. Darwin

My notions sometimes bring good; D^r. Davy has been experimenting at my request, (in order to see how fishes' ova might get transported) on the retention of vitality; & he found that salmon's ova, exposed for 3 whole days to open air, & even some sun-shine, &

they produced fine young fish. D^r. D. has sent a paper to Royal Soc. on the subject.— N.B. Remember to ask about my *distinct* case of "a lady in N. America" who saw fishes' spawn adhering to a Ditiscus [water-beetle].

To J. D. Hooker   13 April [1855]

Down Farnborough Kent.
Ap. 13th.

My dear Hooker

... Thank you very much for the information about the seeds.[1] I had fancied you had some definite opinion that seeds of certain groups could not possibly withstand salt-water. I am not yet prepared to try the experiment on so large a scale as you suggest: indeed I have hardly the means; but I am glad to find I have commenced very much on the principles you suggest, but on a much smaller scale. I have had one experiment some little time in progress, which will I think be interesting, namely seeds in salt water immersed in water of 32°–33°, which I have & shall long have, as I filled a great tank with Snow.— When I wrote last, I was going to triumph over you, for my experiment had in a slight degree succeeded, but this with infinite baseness I did not tell in hopes that you would say that you would eat all the plants, which I could raise after immersion. It is very aggravating that I cannot in the least remember what you did formerly say, that made me think you scoffed at the experiments vastly; for you now seem to view the experiment like a good Christian. I have in small bottles out of doors, exposed to variations of temp., but in shade, exposed to light, as yet only Cress, Radish, Cabbages, Lettuces, Carrots, Celery, & Onion seed; 4 great Families. These after immersion for exactly one week, have all germinated, which I did not in the least expect, (& thought how you w^d. sneer at me) for the water of nearly all & of the cress especially, smelt very badly, & the cress-seed emitted a wonderful quantity of mucus (the Vestiges would have expected them to turn into tadpoles) so as to cohere in a mass; but these seeds germinated & grew splendidly. The germination of all (especially Cress & Lettuces) has been accelerated, except the cabbages, which have come up very irregularly & a good many, I think, dead. One w^d. have thought from native habitat that cabbage w^d. have stood well. The Umbelliferæ & onions seem to stand the salt well. I wash the seed before planting them. I have written to Gardeners' Chronicle; though I doubt whether it was worth while.

If my success seems to make it worth while, I will send a seed list to get you to mark some different classes of seeds. To day I replant the same seeds as above after 14 days immersion. As many sea-current go a mile an hour: even in a week they might be transported 168 miles: the Gulf-stream is said to go 50 & 60 miles a day.— So much & too much on this head; but my geese are always swans. . . .

    Goodbye | My dear Hooker | Most truly yours | C. Darwin

    I plant my salted seeds in glass tumblers (having first tried & recorded rate of germination of same seeds unsalted) so that I can see the seed all the time, before & after germination, on the chimney piece.—

[CD's investigations of the geographical distribution of plants led to a global correspondence with botanists. Asa Gray, professor of botany at Harvard University, became his chief correspondent and authority on the flora of the United States.]

To Asa Gray   25 April [1855]

Down Farnborough Kent
April 25th.

My dear Sir

    I hope that you will remember that I had the pleasure of being introduced to you at Kew. I want to beg a great favour of you, for which I well know I can offer no apology. But the favour will not, I think, cause you much trouble & will greatly oblige me. As I am no Botanist, it will seem so absurd to you my asking botanical questions, that I may premise that I have for several years been collecting facts on "Variation", & when I find that any general remark seems to hold good amongst animals, I try to test it in Plants.—

    I have the greatest curiosity about the alpine Flora of the U.S., & I have copied out of your Manual the enclosed list; now I want to know whether you will be so very kind as to append from memory (I have not for one instant the presumption to wish you to look to authorities) the other habitats or range of these plants: appending "Indig." for such as are confined to the mountains of the U.S.— "Arctic Am." to such as are also found in Arctic America.— "Arctic Eu." to those also found in Arctic Europe:—& "Alps" to those found on any *mountains* of Europe.—"& Arct. Asia" I have compared the list with the plants of Britain, but I am of course afraid of trusting to myself, from ignorance of synonyms &c.—

    I see that there are 22 species common to the White M^{ts} & the M^{ts}. of New York, will you tell me about how wide a space of low land,

on which these alpine plants cannot grow, separates these mountains: I can hardly judge from the height not being marked on the
prolongation of the mountains of Vermont.—

I venture to ask for one more piece of information, viz. whether
you have anywhere published a list of the phanerogamic species common to Europe, as has been done with the shells & Birds, so that a
non-Botanist may judge a little on the relationship of the two floras.
Such a list would be of extreme interest for me in several points of
view & I should think for others. I suppose there would not be more
than a few hundred out of the 2004 species in your Manual. Should
you think it very presumptuous in me to suggest to you to publish
(if not already done) such a list in some Journal?— I would do it
for myself, but I sh^d. assuredly fall into many blunders. I can assure
you, that I perceive how presumptuous it is in me, not a Botanist, to
make even the most trifling suggestion to such a Botanist as yourself;
but from what I saw & have heard of you from our dear & kind
friend Hooker, I hope & think that you will forgive me, & believe
me, with much respect, | Dear Sir | Your's very faithfully | Charles
Darwin

To W. D. Fox   7 May [1855]

Down Farnborough Kent
May 7th—

My dear Fox

. . . I am rather low today about all my experiments,— everything
has been going wrong— the fan-tails have picked the feathers out
of the Pouters in their Journey home— the fish at the Zoological
Gardens after eating seeds would spit them all out again— Seeds
will sink in salt-water— all nature is perverse & will not do as I wish
it, & just at present I wish I had the old Barnacles to work at &
nothing new.—

Well to return to business, nobody, I am sure could fix better for
me, than you, the characteristic age of little chickens: with respect
to skeletons I have feared it w^d be impossible to make them; but I
suppose I shall be able to measure limbs &c by feeling the joints.
What you say about old Cocks just confirms what I thought; & I will
make my skeltons of old cocks.— Sh^d. an old wild Turkey ever die
please remember me: I do not care for Baby turkey. Nor for a mastiff.
Very many thanks for your offer.— I have puppies of Bull-dogs &
Greyhound in salt. —& I have had Carthorse & Race Horse young

colts carefully measured.— Whether I shall do any good I doubt: I
am getting out of my depth.—

Most truly yours. | C. Darwin

To W. D. Fox   17 May [1855]

Down Farnborough Kent
May 17th

My dear Fox.

You will hate the very sight of my handwriting; but after this time
I promise I will ask for nothing more, at least for a long time.— As
you live on Sandy Soil, have you Lizards at all common? If you have,
sh^d. you think it too ridiculous to offer a reward for me for Lizards
eggs to the Boys in your school;—a shilling for every half-dozen, or
more if rare, till you get 2 or 3 dozen & send them to me.— If snake's
eggs were brought in mistake it would be very well, for I want such
also: & we have neither lizards or snakes about here.—

My object is to see whether such eggs will float on sea-water, &
whether they will keep alive thus floating for a month or two in
my cellar. I am trying experiments on transportation of all organic
beings, that I can; & Lizards are found on every isl^d. & therefore I
am very anxious to see whether their eggs stand sea water. Of course
this note need not be answered, without by a strange & favourable
chance you can someday answer it with the eggs.

Your most troublesome friend | C. Darwin

To W. D. Fox   23 May [1855]

Down Farnborough Kent
May 23^d.

My dear Fox

... I had quite forgotten when I wrote to you, that the very common British Lizard is ovo-viviparous! & the chance of getting ova of
the L. agilis, I fear is small. Jersey is evidently the best chance.— I
am going to try land-snail shells & their eggs also. in sea-water.—

To J. D. Hooker   5 June [1855]

Down. Farnborough Kent
June 5th

My dear Hooker

Very many thanks for your seeds & Saxifrage, & such a splendid
lot: ... Thanks, also, for your little note with all the terrible wishes

about the seeds, in which I almost join for I begin to think they are immortal & that the seed job will be another Barnacle job; for I thought the first lot were all dead; & now after 56 days, 6 out of the 7 kinds have come up, though only a few of each.— It was a very good, (though I thought useless at the time) suggestion; to try *cabbage*, *broccoli* & *cawlifower*, the two latter having *everyone* died after 22 days, wheras cabbage itself has germinated well. Having no one to talk to, I must just tell you, what seems to me curious, that the young plants of Tussilago farfara came out of their seeds in the salt-water, & have now kept alive nine days some floating & some at bottom of sea water, & when planted they grow well. ... *Your* lot of seeds have done very badly; partly perhaps owing to their being several of them Greenhouse plants; & partly owing to the seeds being bad; & they are dreadfully slow germinators, which is a great evil, & which no doubt you selected on purpose to vex me.— Miss Thorley & I are doing a *little Botanical work* (!) for our amusement, & it does amuse me very much, viz making a collection of all the plants, which grow in a field, which has been allowed to run waste for 15 years, but which before was cultivated from time immemorial; & we are also collecting all the plants in an adjoining & *similar* but cultivated field; just for the fun of seeing what plants have arrived or dyed out. Hereafter we shall want a bit of help in naming puzzlers.— How dreadfully difficult it is to name plants. ...

I thank you much for Hedysarum: I do hope it is not very precious, for as I told you it is for probably a *most* foolish purpose: I read somewhere that no plant closes its leaves so promptly in darkness, & I want to cover it up daily for $\frac{1}{2}$ hour, & see if I can **teach it** to close by itself, or more easily than at first in darkness. ...

How I do wish I c^d. see you oftener, what good it w^d. do me in my work. But busy as you are, I beg you with *most perfect truth* on no account to trouble yourself in writing often to me, because I write to you.

Good Bye | C. Darwin

I cannot make exactly out why you w^d. prefer continental trans- mission, as I think you do, to carriage by sea: with your general views, I sh^d. have thought you w^d. have been pleased at as many means of transmission as possible.— For my own pet theoretical notions, it is quite indifferent whether they are transmitted by sea or land, as long as some, tolerably probable way is shown. But it shocks my philos- ophy to create land, without some other & independent evidence.

Whenever we meet, by a very few words I sh^d. I think more clearly understand your views. . . .

I have just made out my first Grass, hurrah! hurrah!² I must confess that Fortune favours the bold, for as good luck w^d have it, it was the easy Anthoxanthum odoratum: nevertheless it is a great discovery; I never expected to make out a grass in all my life. So Hurrah. It has done my stomach surprising good.—

To Asa Gray   8 June [1855]

Down Farnborough Kent
June 8th

My dear Sir

I thank you cordially for your remarkably kind letter of the 22^d ult°., & for the extremely pleasant & obliging manner in which you have taken my rather troublesome questions. I can hardly tell you how much your list of Alpine plants has interested me, & I can now in some degree picture to myself the plants of your alpine summits. The new Edit. of your Manual is *capital* news for me: I know from your preface how pressed you are for room, but it would take no space to append (Eu.) in bracket to every European plant, & as far as I am concerned this would answer every purpose. From my own experience whilst making out English plants in our Manuals, it has often struck me, how much interest it would give if some notice of their range had been given, & so I cannot doubt your American enquirers, & beginners w^d. much like to know which of their plants were indigenous, & which European.

Would it not be well in the Alpine plants to append the very same additions which you have now sent me in M.S; though here, owing to your kindness, I do not speak selfishly, but merely pro bono Americano publico.— I presume it w^d. be too troublesome to give in your Manual the habitats of those plants found West of Rocky mountains; & likewise those found in Eastern Asia, taking the Yenesei (?) which, if I remember right according to Gmelin is the main partition line of Siberia. Perhaps Siberia more concerns the northern Flora of N. America. The ranges of the plants, to the East & West, viz whether most found are in Greenland & Western Europe, or in E. Asia appears to me a very interesting point as tending to show whether the migration has been Eastward or Westward.— Pray believe me, that I am most entirely conscious that the *only use* of these remarks is to show a Botanist what points a non-Botanist is

142

curious to learn; for I think everyone who studies profoundly a subject often becomes unaware what points the ignorant require information. I am so very glad that you think of drawing up some notice on geographical distribution, for the area of the Manual strikes me as in some points better adapted for comparison with Europe than that of the whole of N. America.—

You ask me to state definitely some of the points on which I much wish for information; but I really hardly can, for they are so vague, & I rather wish to see what results will come out from comparisons, than have as yet defined objects. I presume that like other Botanists you would give for your area, the proportions (leaving out introduced plants) to the whole of the great leading families: this is one point, I had intended (& indeed have done roughly) to tabulate from your Book, but of course I could have done it only *very imperfectly*. I should, also, of course have ascertained the proportion to the whole Flora of the European plants (leaving out introduced) *& of the separate great families*, in order to speculate on means of transportal. By the way I ventured to send a few days ago a copy of the Gardeners' Chronicle, with a short report by me of some trifling experiments which I have been trying on the power of seeds to withstand sea-water. I do not know, whether it has struck you, but it has me, that it would be adviseable for Botanists to give in *whole numbers*, as well as in the lowest fraction, the proportional numbers of the Families.— thus I make out from your Manual that of the *indigenous* plants the proportion of the Umbelliferæ are $\frac{36}{1798} = \frac{1}{49}$; for without one knows the *whole* numbers, one cannot judge how really close the numbers of the plants of the same family are in two distant countries; but very likely you may think this superfluous.— mentioning these proportional numbers, I may give as an instance of the sort of points, **& how vague & futile they often are** which I *attempt* to work out, that reflecting on R. Brown & Hooker's remark, that near identity of proportional number of the great Families, in two countries, shows probably that they were once continuously united, I thought I would calculate the proportions, of, for instance, the *introduced* Compositæ in Grt. Britain to **all** the introduced plants, & the result was $\frac{10}{92} = \frac{1}{9.2}$. In our *aboriginal* or indigenous flora the proportion is 1/10; & in many other cases I found an equally striking correspondence: I then took your Manual, & worked out the same question; here I found in the Compositæ an almost equally striking correspondence, viz $\frac{24}{206} = \frac{1}{8}$ in the introduced plants, and $\frac{223}{1798} = \frac{1}{8}$ in the indigenous; but when I came to the other

Families, I found the proportions entirely different showing that the coincidences in the British Flora were probably accidental!—

You will, I presume, give the proportion of the species to the genera, ie show on an average how many species each genus contains; though I have done this for myself.—

If it would not be too troublesome do you not think it w^d be very interesting, & give a very good idea of your Flora, to divide the species into 3 groups, viz (a) species common to the old word, stating numbers common to Europe & Asia (b) indigenous species, but belonging to genera found in the old world, & (c) species belonging to genera confined to America or the New World. To make, (according to my ideas perfection perfect) one ought to be told whether there are other cases like Erica of genera common in Europe or in old world not found in your area.—

But honestly I feel that it is quite ridiculous my writing to you at such length on such subject, but as you have asked me, I do it gratefully, & write to you, just as I should to Hooker, who often laughs at me unmercifully, & I am sure you have better reason to do so.—

To J. D. Hooker 5 July [1855]

Down
July 5th

My dear Hooker

I sh^d be a much happier man if you would specifically name this grass for me: it has fairly beaten me: I am not sure even whether it is a Festuca; I feel pretty sure that it is not F. pratensis.— It grows on rather dry chalky banks. I have collected 35 species & named all, I believe, correctly excepting this & one other, which latter I think I shall make out, when in fuller flower.—

Ever yours | C. Darwin

P.S. | *Lots* of Celery, & some Onions, & Carrotts & even one Radish & one Lettuce have come up after 85 days in salt-water!—

To Asa Gray 24 August [1855]

Down Farnborough Kent
Aug. 24th

My dear Dr Gray

... I really hardly know how to thank you enough for the *very* great trouble which the list of "close species" must have caused you.— What knowledge & labour & judgment is condensed in that little

144

sheet of note-paper! I fear that you will think the object not at all worth the labour; but I can only say that if I could have done it myself, I would have done it, had it caused me ten times the labour which it must have caused you. I had met with a remark by Fries that the species of large genera are more closely related to each other, than are the species of small genera. I consulted a very good entomologist, & Hooker & Bentham, & they did not at all believe in this. But several facts & considerations, nevertheless, made me think that there might be some truth in it; and all general statements of such kind it is my object in my eclectic, peddling sort of work to test. It occurred to me that if I could get some good systematists, not species-splitters, to mark (without the object being known) the close species in a list; then if I counted the average number of the species in such genera, & compared it with the general average (for this end all the genera with *single* species have to be omitted; & I have omitted Salix & Carex also) of the species to the genera in the same country; it would, *to a certain extent*, tell whether on average the close species occurred in the larger genera.

Now in your M.S list (Salix & Carex being omitted) there are 115 genera & these have 6.37 species to genus: whereas in your Manual (omitting Salix & Carex & all genera with *single* species) the average is 4.67. So that it seems that when *many* organic forms are allied, making what is called a genus, some of them are apt to be more closely allied than are the species in the smaller genera. M^r. H. C. Watson has marked for me the British Flora, & the same result is given. I know how vague all such results must be, & there may be some fallacy (should the fallacy be apparent to you, I sh^d be most grateful to be informed.) in the result, but I cannot detect it; & am inclined to believe that the above proposition may be trusted; but I shall of course try to test it by other means.—

Pray accept my true & cordial thanks for all your very great kindness, & believe me, Your's very truly | Ch. Darwin ...

[In the summer of 1855, CD met William Bernhard Tegetmeier, a leading authority on poultry. Their correspondence lasted until 1881.]

To William Bernhard Tegetmeier 31 August [1855]

Down Farnborough Kent
Aug. 31st.

Dear Sir

I have been thinking over your offer of helping me to the dead bodies of some of the good birds of Poultry.— Really considering how complete a stranger I am to you, I think it one of the most goodnatured offers ever made to me.— I have hardly the means to keep all the kinds of poultry, & to buy *first-rate* birds, merely to make skeletons of them, I should think too great an outlay. Therefore if you can help me even to a few it would be a very great assistance.

I have thought it would be best to enclose a list, but pray do not for a minute suppose that I am so unreasonable as to imagine that you can take the trouble to supply me with nearly all; but even a few would be of great service to me. Forgive me for adding that I hope that you will be so good as to remember whatever expence you may be put to for carriage, porterage, booking, baskets &c, & allow me to repay you. The trouble is *very much* more than I could have expected you to put yourself to, & I should be ashamed of myself if in addition, you were put even to a trifling expence. I do not think I shall be in London very soon, but when I am I will propose to call for an hour if you should chance to be disengaged.—

I am sure I have cause to offer many apologies, & beg to remain Dear Sir | Your's truly obliged | Ch. Darwin

I published some years since a Natural History Journal of my Travels, which has been liked by some naturalists: if you should feel the least interest in seeing it, I sh^{d} be proud to present you with a copy.—

To *Gardeners' Chronicle*　21 November [1855]

As you have published notices by Mr. Berkeley and myself on the length of time seeds can withstand immersion in sea-water, you may perhaps like to hear, without minute details, the final results of my experiments. The seed of Capsicum, after 137 days' immersion, came up well, for 30 out of 56 planted germinated, and I think more would have grown with time. Of Celery only 6 out of some hundreds came up after the same period of immersion. One single Canary seed grew after 120 days, and some Oats half germinated after 120; both Oats and Canary seed came up pretty well after only 100 days. Spinach germinated well after 120 days. Seed of Onions, Vegetable Marrow, Beet, Orache and Potatoes, and one seed of Ageratum mexicanum grew after 100 days. A few, and but very few, seed of Lettuce, Carrot, Cress, and Radish came up after 85 days' immersion. It is remarkable how differently varieties of the same species have withstood the

ill effects of the salt water; thus, seed of the "Mammoth White Broccoli" came up excellently after 11 days, but was killed by 22 days'
immersion; "early Cauliflower" survived this period, but was killed
by 36 days; "Cattell's Cabbage" survived the 36 days, but was killed
by 50 days; and now I have seed of the wild Cabbage from Tenby
growing so vigorously after 50 days, that I am sure that it will survive
a considerably longer period. But the seed of the wild Cabbage was
fresh, and some facts show me that quite fresh seed withstands the
salt water better than old, though very good seed. With respect to an
important point in my former communication of May 26th, permit
me to cry *peccavi;* having often heard of plants and bushes having
been seen floating some little distance from land, I assumed—and
in doing this I committed a scientific sin—that plants with ripe seed
or fruit would float at least for some weeks. I always meant to try
this, and I have now done so with sorrowful result; for having put in
salt-water between 30 and 40 herbaceous plants and branches with
ripe seed of various orders, I have found that all (with the exception
of the fruit of evergreens)[3] sink within a month, and most of them
within 14 days. So that, as far as I can see, my experiments are of
little or no use (excepting perhaps as negative evidence) in regard to
the distribution of plants by the drifting of their seeds across the sea.
Can any of your readers explain the following sentence by Linnæus,
pointed out to me by Dr. Hooker, "Fundus maris semina non de-
struit"? Why does Linnæus say that the bottom of the sea does not
destroy seeds? The seeds which are often washed by the Gulf Stream
to the shores of Norway, with which Linnæus was well acquainted,
float, as I have lately tried. Did he imagine that seeds were drifted
along the bottom of the ocean? This does not seem probable, from
the currents of the sea, at least many of them, being superficial.
Charles Darwin, Down, Nov. 21. . . .

To Thomas Campbell Eyton   26 November [1855]

Down Bromley Kent
Nov. 26th.

My dear Eyton

As you have had such great experience in making skeletons, will
you be so kind as to take the trouble to give me some pieces of
information. But I must premise that I have been making a few, &
when I took the body out of the water, the smell was so dreadful that
it made me reach awfully. Now I was told that if I hung the body of

a bird or small quadruped up in the air & allowed the flesh to decay off, & the whole to get dry, that I could boil the mummy in water with caustic soda, & so get it nearly clean, but not white, with very little smell. What do you think of this plan? And pray tell me how do you get the bones moderately clean, when you take the skeleton out, with some small fragments of putrid flesh still adhering. It really is most dreadful work.— Lastly do you pluck your Birds?—

I am getting on with my collection of Pigeons, & now have pairs of ten varieties alive & shall on Saturday receive two or three more kinds.—

Do pray help me with your advice, & forgive this trouble.

Your's very truly | C. Darwin

T. C. Eyton Esq^{e}

1856

To J. E. Gray   14 January [1856]

Down Bromley Kent
Jany. 14th

My dear Gray

You have often helped me, will you be so kind as to help me this time in regard to the enclosed memorandum, with M^r. Birch. It is my only imaginable channel by which I can ever learn anything about the varieties of our domesticated animals & plants in China.— Do pray use your interest for me with M^r. Birch; I could not ask myself.—

My dear Gray | Yours very truly | C. Darwin

[Enclosure]

Is there any translation of any Chinese work, ancient or modern, descriptive, or even simply enumerative, of the varieties of *domestic* Pigeons & Fowls or Ducks kept by the Chinese; & likewise of the Dogs, sheep, cattle &c; but I care more about the former even than the latter.— And the same in regard to the varieties of cultivated plants, but more especially of tobacco & maize; for these latter plants, the work, of course, must not be ancient.—

If any such Chinese agricultural work or Encyclopædia exists in the British Museum but has not been translated, would it be possible for Mr Birch, & would he be so very kind as to take the trouble as to look at it (& as probably saving him a little trouble) & let me be present to note down names of any varieties mentioned, if such are specified.. This would be of extreme interest to me; but I hardly know how great a favour I am asking, for Chinese seems to be so wonderfully difficult to read.—

C. Darwin

To W. B. Tegetmeier   14 January [1856]

Down Bromley Kent
Jan. 14th.—

My dear Sir

I have been unwell for a week, otherwise I sh^d. not have left so many days elapse without thanking you very sincerely for your most kind offer of buying for me old Cocks at Stephens.— I have only *one* skeleton as yet, of a good Spanish Cock, so that I sh^d be glad of anything or everything, which you consider a distinct breed. I sh^d. be willing to go to 5^s per bird.— My old friend the Rev. R. Pulleine (whose name, I daresay you have heard as a good Poultry judge) sent me a message the other day that he was sure that Mr Baily would at his request send me anything; but I believe your scheme is more sure & I will not as yet try Baily. I am in no hurry. If I succeed in my attempts to get the *skins* of Poultry from all quarters of the world, I shall want skins of the breeds of England for comparison; so if you stumble on a bird *in good plumage*, I wish you would have its neck broken, instead of cut, & then I shall understand that you think it worth skinning, instead of skeletonising. Should I *ultimately* succeed in making good collection of skins & skeletons of our domestic birds, I shall give whole to British Museum. ...

With very sincere thanks | Your's truly | C. Darwin ...

To W. E. Darwin   [26 February 1856]

[Down]
Tuesday Evening

My dear old Willy

I was very glad to get your letter this morning, but I wish I could hear that your leg was quite healed: be sure tell us particularly how it goes on.— I am glad to hear of your sixth-form power; it is good to get habit of command & discretion in commanding; & you unfortunate wretch, how you will enjoy reading the prayers, & keeping the accounts; as for carving you will cut a good figure.— You know Mamma is at Hartfield with the 3 little chaps; I enclose a note from Lenny. He sent such a funny one lately to Leith Hill: it began "Baby has a shag coat, but it is brown.— I have bought some sealing wax & I have bought some note paper: it is quite true.— Is not this a jolly letter?." & so on for 4 pages.— Snow, the dog has come back, very fat & is just as much at home as before.—

We have today cut down & grubbed the big Beech tree by the roundabout: I find by the rings it is 77 years old: I am going to try whether there are any seeds in the earth from right under it, for they must have been buried for 77 years.— I am getting on splendidly with my pigeons; & the other day had a present of Trumpeters, Nuns & Turbits; & when last in London, I visited a jolly old Brewer, who keeps 300 or 400 most beautiful pigeons & he gave me a pair of pale brown, quite small German Pouters: I am building a new house for my tumblers, so as to fly them in summer.— I am sorry to say that I have had to strike out your name for Athenæum Club, as you cannot be entered till 18 years old. Several members mistook you for me & Lord Overstone called here to say that he should propose me to be elected by the Committee, who have power of electing 8 members every year, so that I have had a deal of bother on the subject.— I sh^d. like to hear what you do in Chemistry.— Good night, my dear old man,

Your affect. father | C. D.— ...

To W. D. Fox   15 March [1856]

Down Bromley Kent

March 15th

My dear Fox.

... Many thanks for your continued remembrance of me & my poultry skeletons: I am making some progress & have been working a little at their ancient History & was yesterday in the British Museum getting old Chinese Encyclopedias translated. This morning I have been carefully examining a splendid Cochin Cock sent me (but I sh^d. be glad of another specimen) & I find several important differences in number of feathers in alula, primaries & tail, making me suspect quite a distinct species.— I am getting on best with Pigeons, & have now almost every breed known in England alive: I shall find, I think great differences in skeleton for I find extra rib & *dorsal* vertebra in Pouter.—

I have just ordered the Cottage Gardener: Mr Tegetmeier is a very kind & clever little man; but he was not authorised to use my name in any way, & we cannot be said to be working at all together; for our objects are very different, & he began on skulls before I had thought on subject: I have not yet looked at our pickled chickens & hardly know when I shall, for I have my hands very full of work; but they will come in some day most useful, as will a large series of young Pigeons, which I have myself killed & pickled.—

I sh^d be very glad of old Sebright Bantam. ...

How I do wish I had you nearer to talk over & benefit by your opinions on the many odds & ends on which I am at work. Sometimes I fear I shall break down for my subject gets bigger & bigger with each months work.—

My dear old friend | Most truly yours | Ch. Darwin

[During a visit to Down in April, Lyell had heard a detailed account of CD's species theory and had urged CD to publish an account of his work in order to establish his priority.]

To Charles Lyell   3 May [1856]

Down Bromley Kent
May 3^{d}.

My dear Lyell

... It is really striking (but almost laughable to me) to notice the change in Hookers & Huxley's opinions on species during the last few years.—

With respect to your suggestion of a sketch of my view; I hardly know what to think, but will reflect on it; but it goes against my prejudices. To give a fair sketch would be absolutely impossible, for every proposition requires such an array of facts. If I were to do anything it could only refer to the main agency of change, selection,—& perhaps point out a very few of the leading features which countenance such a view, & some few of the main difficulties. But I do not know what to think: I rather hate the idea of writing for priority, yet I certainly sh^{d}. be vexed if any one were to publish my doctrines before me.— Anyhow I thank you heartily for your sympathy. I shall be in London next week, & I will call on you on Thursday morning for one hour precisely so as not to lose much of your time & my own: but will you let me this one time come as early as 9 oclock, for I have much which I must do, & the morning is my strongest time.

Farewell | My dear old Patron | Yours | C. Darwin

By the way *three* plants have now come up out of the earth **perfectly** enclosed in the roots of the trees.— And 29 plants in the table-spoon-full of mud out of little pond: Hooker was surprised at this, & struck with it, when I showed him how much mud I had scraped off one Duck's feet.—

If I did publish a short sketch, where on earth should I publish it? ...

To J. D. Hooker   9 May [1856]

Down Bromley Kent
May 9th

My dear Hooker

... With respect to Huxley, I was on point of speaking to Crawfurd & Strezlecki (who will be on committee of Athenæum) when I bethought me of how Owen would look & what he would say.

Cannot you fancy him, with a red face, dreadful smile & slow & gentle voice, asking, "Will Mr Crawfurd tell me what Mr Huxley has done, deserving this honour; I only know that he differs from, & disputes the authority of Cuvier, Ehrenberg & Agassiz as of no weight at all".— And when I began to consider what to tell Mr Crawfurd to say, I was puzzled, & could refer him only to some *excellent* papers in R. Trans. for which the medal had been awarded. But I doubt *with an opposing faction*, whether this would be considered enough, for I believe real scientific merit is not thought enough, without the person is generally well known; now I want to hear what you *deliberately* think on this head: it would be bad to get him proposed & then rejected; & Owen is very powerful.—

Lastly, & of course especially, about myself; I very much want advice & *truthful* consolation if you can give it. I had good talk with Lyell about my species work, & he urges me strongly to publish something. I am fixed against any periodical or Journal, as I positively will *not* expose myself to an Editor or Council allowing a publication for which they might be abused.

If I publish anything it must be a *very thin* & little volume, giving a sketch of my views & difficulties; but it is really dreadfully unphilosophical to give a resumé, without exact references, of an unpublished work. But Lyell seemed to think I might do this, at the suggestion of friends, & on the ground which I might state that I had been at work for 18 years, & yet could not publish for several years, & especially as I could point out difficulties which seemed to me to require especial investigation. Now what think you?. I sh^d. be really grateful for advice. I thought of giving up a couple of months & writing such a sketch, & trying to keep my judgment open whether or no to publish it when completed. It will be simply impossible for me to give exact references; anything important I sh^d. state on authority of the author generally; & instead of giving all the facts on which I ground any opinion, I could give by memory only one or two. In Preface I would state that the work could not be considered strictly scientific, but a mere sketch or outline of future work in which full references &c sh^d. be given.— Eheu, eheu, I believe I sh^d. sneer at anyone else doing this, & my only comfort is, that I *truly* never dreamed of it, till Lyell suggested it, & seems deliberately to think it adviseable.

I am in a peck of troubles & do pray forgive me for troubling you.—

Yours affecti^y | C. Darwin ...

To J. D. Hooker 11 May [1856]

Down Bromley Kent
May 11th

My dear Hooker

... I am extremely glad you think well of a separate "Preliminary Essay" i.e. if anything whatever is published; for Lyell seemed rather to doubt on this head; but I cannot bear the idea of *begging* some Editor & Council to publish & then perhaps to have to *apologise* humbly for having led them into a scrape. In this one respect I am in the state, which according to a very wise saying of my Father's, is the only fit state for asking advice, viz with my mind firmly made up, & then, as my Father used to say, *good* advice was very comfortable & it was easy to reject *bad* advice.— But Heaven knows I am not in this state with respect to publishing at all any preliminary essay. It yet strikes me as quite unphilosophical to publish results without the full details which have led to such results.

It is a melancholy, & I hope not quite true view of your's that facts will prove anything, & are therefore superfluous! But I have rather exaggerated,, I see, your doctrine. I do not fear being tied down to error, i.e. I feel pretty sure I should give up anything false published in the preliminary essay, in my larger work; but I may thus, it is very true, do mischief by spreading error, which as I have often heard you say is much easier spread than corrected. I confess I lean more & more to at least making the attempt & drawing up a sketch & trying to keep my judgment whether to publish open. But I always return to my fixed idea that it is dreadfully unphilosophical to publish without full details. I certainly think my future work in full would profit by hearing what my friends or critics (if reviewed) thought of the outline.—

To anyone but you I sh^d. apologise for such long discussion on so personal an affair; but I believe, & indeed you have proved it by the trouble you have taken, that this would be superfluous.

Your's truly obliged | Ch. Darwin ...

P.S | What you say (for I have just reread your letter) that the Essay might supersede & take away all novelty & value from my future larger Book, is very true; & that would grieve me beyond everything. On the other hand, (again from Lyell's urgent advice) I published a preliminary sketch of Coral Theory & this did neither good nor harm.— I begin *most heartily* to wish that Lyell had never put this idea of an Essay into my head.

To J. D. Hooker   21 [May 1856]

Down Bromley Kent
21st

My dear Hooker

I have got the Lectures & have read them. The Lectures strike me as very clever. Though I believe, as far as my knowledge goes that Huxley is right, yet I think his tone very much too vehement, & I have ventured to say so in a note to Huxley.—[1] I had not thought of these Lectures in relation to the Athenæum, but I am inclined quite to agree with you & that we had better pause before anything is said. It might be urged as a real objection the way our friend falls foul of every one (N.B I found Falconer very indignant at the manner in which Huxley treated Cuvier in his R. In^{n}. Lecture; & I have gently told Huxley so.) I think we had better do nothing, to try in earnest to get a great Naturalist into Athenæum & fail, is far worse than doing nothing—

How strange, funny & disgraceful that nearly all—(Faraday, Sir J. Herschel at least exceptions) our great men are in quarrels in couplets; it never struck me before.— ...

Ever yours | C. Darwin ...

To Charles Lyell   16 [June 1856]

Down
16th.—

My dear Lyell

I am going to do the most impudent thing in the world. But my blood gets hot with passion & runs cold alternately at the geological strides which many of your disciples are taking.

Here, poor Forbes made a continent to N. America & another (or the same) to the Gulf weed.— Hooker makes one from New Zealand to S. America & round the world to Kerguelen Land. Here is Wollaston speaking of Madeira & P. Santo "as the sure & certain witnesses" of a former continent. Here is Woodward writes to me if you grant a continent over 200 or 300 miles of ocean-depths (as if that was nothing) why not extend a continent to every island in the Pacific & Atlantic oceans!

And all this within the existence of recent species! If you do not stop this, if there be a lower region for the punishment of geologists, I believe, my great master, you will go there. Why your disciples in a slow & creeping manner beat all the old catastrophists who ever lived.— You will live to be the great chief of the catastrophists!

There, I have done myself a great deal of good & have exploded my passion.

So my master forgive me & believe me | Ever yours | C. Darwin . . . Dont answer this, I did it to ease myself.—

To J. D. Hooker 17–18 [June 1856]

Down Bromley Kent
17th

My dear Hooker

I was actually wishing *much* to hear what you thought on the two subjects, to which your note, received this morning is chiefly devoted.— I did not like to give up the time to form a very certain judgment to my own satisfaction, in Falconer v. Huxley. But the article struck me as very clever.— I rather lean to the Huxley side, & without Falconer can say that he could have told, without the knowledge of habits of any bears, that the Polar bear was carnivorous & the brown bear frugivorous from structure alone, I think Huxleys argument best.— But to deny all reasoning from adaptation & so called final causes, seems to me preposterous. But I am most heartily sorry at the whole dispute: it will prevent two very good men from being friends. . . .

I have been very deeply interested by Wollaston's book, though I differ *greatly* from many of his doctrines. Did you ever read anything so rich, considering how very far he goes, as his denunciations against those who go further, "most mischievous" "absurd", "unsound". Theology is at the bottom of some of this. I told him he was like Calvin burning a heretick.— It is a very valuable & clever book in my opinion.— He has evidently read very little out of his own line: I urged him to read the New Zealand Essay. His Geology also is rather *eocene*, as I told him. In fact I wrote most frankly; I fear too frankly; he says he is sure that *ultra*!-honesty is my Characteristic: I do not know whether he meant it as a sneer; I hope not.—

Talking of eocene geology, I got so wrath about the Atlantic continent, more especially from a note from Woodward (who has published a **capital** book on shells) who does not seem to doubt that *every island* in Pacific & Atlantic are the remains of continents, submerged within period of existing species; that I fairly exploded & wrote to Lyell to protest & summed up all the continents created of late years by Forbes, (the head *sinner*!) *yourself*, Wollaston, & Woodward & a pretty nice little extension of land they make altogether! I am fairly *rabid* on the question & therefore, if not wrong already, am pretty sure to become so. ...

    Adios. | C. Darwin ...

To Charles Lyell 8 July [1856]

Down Bromley Kent
July 8th.

My dear Lyell

Very many thanks for your two notes & especially for Maury's map: also for Books which you are going to lend me.

I am sorry you cannot give any verdict on continental extensions; & I infer that you think my arguments of not much weight against such extensions: I know I wish I could believe.—

I have been having a good look at Maury (which I once before looked at) & in respect to Madeira & co, I must say that the chart seems to me against land-extension explaining introduction of organic beings. Madeira, the Canaries & Azores are so tied together that I sh^d. have thought that they ought to have been connected by some bank if changes of level had been connected with their organic relation. The azores ought too to have shown, more connection with America. I had sometimes speculated whether icebergs could account for the greater number of European plants & their more northern character on the Azores compared with Madeira; but it seems dangerous until boulders are found there.

One of the most curious points in Maury, as it strikes me, is the little change, which about 9000 feet of sudden elevation would make in the continent visible, & what a prodigious change 9000 feet subsidence would make! Is this difference due to denudation during elevation? Certainly 12,000 feet elevation would make a prodigious change.—

I have just been quoting you in my essay on ice carrying seeds in S. hemisphere; but this will not do in all the cases.— I have had a week of such labour in getting up the relations of all the antarctic floras from Hooker's admirable works. Oddly enough I have just finished in great detail giving evidence of coolness in Tropical regions during the glacial epoch, & the consequent migration of organisms through the Tropics. There are a good many difficulties, but upon the whole it explains much. This has been a favourite notion with me almost since I wrote on erratic boulders of the south.— It harmonises with the modification of species, & without admitting this awful postulate the glacial epoch in the south & Tropics does not work in well. About Atlantis, I doubt whether the Canary islands are as much more related to the continent as they ought to be if formerly connected by continuous land.—

Yours most truly | C. Darwin

Hooker with whom I have formerly discussed the notion of the world or great belts of it having been cooler, though he at first saw great difficulties (& difficulties there are great enough) I think is much inclined to adopt the idea. With modification of specific forms it explains some wondrous odd facts in distribution.

But I shall never stop if I get on this subject, on which I have been at work, sometimes in triumph & sometimes in despair, for the last month.

To J. D. Hooker   19 July [1856]

Down Bromley Kent
July 19th

My dear Hooker

I thank you warmly for the *very kind* manner with which you have taken my request. It will in truth be a most important service to me; for it is absolutely necessary that I sh^d. discuss single & double creations, as a very crucial point on the general origin of species, & I must confess, with the aid of all sorts of visionary hypotheses, a very hostile one.—[2]

I am delighted that you will take up possibility of crossing; no Botanist has done so which I have long regretted … I am far from expecting that no cases of *apparent* impossibility will be found; but certainly I expect that ultimately they will disappear; for instance Campanulaceæ seemed a strong case, but now it is pretty clear that they must be liable to crossing. Sweet Peas—Bee-orchis, & perhaps Hollyocks are, at present, my greatest difficulties; & I find I cannot experimentise by castrating Sweet Peas, without doing fatal injury. Formerly I felt most interest on this point as one chief means of eliminating varieties; but I feel interest now in other ways.—

One *general* fact makes me believe in my doctrine, is that **no** terrestrial animal in which semen is liquid is hermaphrodite except with mutual copulation: in terrestrial plants in which the semen is dry there are many hermaphrodites.— Indeed I do wish I lived at Kew or at least so that I could see you oftener.

If you were to take up crossing, I w^d. look over my notes, which perhaps w^d. guide you to some of the most difficult cases. . . .
My dear Hooker | Yours affectly | C. Darwin . . .

To T. C. Eyton   31 August [1856]

Down Bromley Kent
Aug. 31st

Dear Eyton

I thank you heartily for your note & for your promise of more information on Pigs, about which I am very curious.— By the way Bechstein asserts that the number of incisors varies greatly in domestic pigs:[3] I am myself going to collect Pigs jaws (no other part) to see whether he is to be trusted. Have you ever noticed this? I sh^d like to confirm Bechstein on your authority. . . .

One of the subjects which gives me most trouble for my work, is means of distribution in the case of species found on distant islands; I have lately been trying the powers of resistance of seeds to sea-water,—their powers of floating—the number of living seeds in earth & mud &c &c.— Would you render me a little assistance in this line? My walking days are over, never to return. I want to know whether on a wet muddy day, whether birds feet are dirty: I am going to send my servant out with some keeper & he shall wash all the partridges feet & save the dirty water!!

But I want especially to know whether herons or any waders (we have no ponds hereabouts) or water-birds when suddenly sprung have *ever* dirty feet or beaks? I found in 2 large table-spoon full of mud from a little pond from beneath the water 53 plants germinated.—

Do you know when owl or Hawk eats a little bird, how soon it throws up pellet? Can it throw up pellet whilst on wing? How I sh^d. like to get a collection of pellets & see whether they contained any seeds capable of germination. Could your gamekeepers find a roosting place, & collect a lot for me?—

Lastly (if you are not sick of my enquiries) have you ever examined the stomachs of dace & other white fish? Do they ever eat seeds; I know it is good to bait a place with grains. For like the house which Jack built, a heron might eat a fish with seed of water plant & then fly to another pond.

I have been trying for a year with no success to get some dace &c. Have you any & could you catch some in net. & order your kitchen maid to clean them, & you c^d. send me the whole stomach & I would sow the contents on burnt earth with every proper precaution. If ever your goodnature sh^d lead you to send me any such rubbish; it might be put in bladder or tin foil & sent by Post, & if you will not think me very impertinent I could repay you the shilling or two for postage; as the rubbish w^d thus come much quicker & cheaper to me.

Do you mean to collect cats' skeletons: Sir C. Lyell has odd Persian & I have *heard* of another odd cat & I w^d request their carcases to be sent to you, if you cared about them. But I fancy cats are much mixed beings.—

Well I have put your words, that you like hearing from old naturalist friends, to a severe test. So forgive me & believe me, | Your's most truly | Ch. Darwin

To J. D. Dana 29 September [1856]

Down Bromley Kent
Sept. 29th

My dear Sir

... I am working very hard at my subject of the variation & origin of species, & am getting M.S. ready for press, but when I shall publish, Heaven only knows, not I fear for a couple of years but whenever I do the first copy shall be sent to you.— I have now been for 19 years with this subject before me; but it is too great for me, especially as my memory is not good. I have of late been chiefly at work on domestic animals, & have now got a considerable collection of skeletons: I am surprised how little this subject has been attended to: I find very grave differences in the skeletons for instance of domestic rabbits, which I think have all certainly descended from one parent wild stock. But Pigeons offer the most wonderful case of variation, & as it seems to me conclusive evidence can be offered that they are all descended from C. livia.

In the case of Pigeons, we have (& in no other case) we have much *old* literature & the changes in the varieties can be traced. I have now a grand collection of living & dead Pigeons; & I am hand & glove with all sorts of Fanciers, Spital-field weavers & all sorts of odd specimens of the Human species, who fancy Pigeons.—

I know that you are not a believer in the doctrine of single points of creation, in which doctrine I am strongly inclined to believe, from *general* arguments; but when one goes into detail there are certainly **frightful** difficulties. No facts seem to me so difficult as those connected with the dispersal of Land Mollusca. If you ever think of, or hear of, any odd means of dispersal of any organisms I sh^{d}. be *infinitely* obliged for any information; as no one subject gives me such trouble as to account for the presence of the same species of terrestrial productions on oceanic islands; for I cannot swallow the prevalent fashion in England of believing that all islands within recent times have been connected with some continent.—

You will be rather indignant at hearing that I am becoming, indeed I sh^{d}. say have become, sceptical on the permanent immutability of species: I groan when I make such a confession, for I shall have little sympathy from those, whose sympathy I alone value.— But anyhow I feel sure that you will give me credit for not having come to so heterodox a conclusion, without much deliberation. How (I think) species become changed I shall explain in my Book, but my views are very different from those of that clever but shallow book, the Vestiges.—

It is my intention to give fully all the facts in favour of the eternal immutability of species & I have taken as much pains to collect them, as I possibly could do. But what my work will turn out, I know not; but I do know that I have worked hard & honestly at my subject.

Agassiz, if he ever honours me by reading my work, will throw a boulder at me, & many others will pelt me; but magna est veritas &c, & those who write against the truth often, I think, do as much service as those who have divined the truth; so that if I am wrong I must comfort myself with this reflection. It may sound presumptious, but I think I have to a certain extent staggered even Lyell.—

But I am scribbling (in a very bad handwriting moreover) in a shameful manner all about myself; so I will stop with cordial good wishes for yourself & family, & pray believe me, my dear Sir | Your sincere & heteredox friend | Ch. Darwin

We are all here now much interested in American politics— You will think us **very impertinent**, when I say how fervently we wish you in the North to be free.—

P.S. | I have long thought that geologists not having found this or that form in this or that formation was *very poor* evidence of such forms not having then existed; in this respect differing, as wide as the poles, from the great Agassiz, who seems to me to retreat a step & take up a new position with a front so bold as to be admirable in a soldier.— Well, in case of cirripedes I thought, as stated in Preface in my Fossil Lepadidæ, that the evidence was so good, that I did believe that no *Sessile* cirripede existed before the Tertiary period.

But yesterday I received from M. Bosquet of Maestricht a beautiful drawing of **perfect** Chthamalus from the Chalk!!—
Never again will I put any trust in negative geological evidence.—

To J. D. Hooker   11–12 November [1856]

Down Bromley Kent
Nov. 11th.—

My dear Hooker

I thank you more *cordially* that you will think probable, for your note. Your verdict has been a great relief.—⁴ On my honour I had no idea whether or not you would say it was (& I knew you would say it very kindly) so bad, that you would have begged me to have burnt the whole. To my own mind my M.S relieved me of some few difficulties, & the difficulties seemed to me pretty fairly stated, but I had become so bewildered with conflicting facts, evidence, reasoning & opinions, that I felt to myself that I had lost all judgment.— Your general verdict is *incomparably* more favourable than I had anticipated.

Very many thanks for your invitation: I had made up my mind on my poor wifes account not to come up to next Phil. Club; but I am so much tempted by your invitation, & my poor dear wife is so goodnatured about it, that I think I shall not resist, ie if she does not get worse.— I w^d. come to dinner at about same time as before, if that w^d suit you & I do not hear to contrary, & w^d. go away by the early train ie about 9 olock.— I find my present work tries me a good deal & sets my heart palpitating, so I must be careful.— But I sh^d. so much like to see Henslow, & likewise meet Lindley if the fates will permit. You will see, whether there will be time for any criticism in detail on my M.S. before dinner. Not that I am in the *least* hurry, for it will be months before I come again to Geograph. Distrib.; only I am afraid of your forgetting any remarks.—

I do not know whether my very trifling observations on means of distribution are worth your reading, but it amuses me to tell them.

The seeds which the Eagle had in stomach for 18 hours looked so fresh that I would have bet 5 to 1 they would all have grown; but some kinds were **all** killed & 2 oats 1 Canary seed, 1 Clover & 1 Beet alone came up! now I sh^d. have not cared swearing that the Beet w^d. not have been killed, & I sh^d have fully expected that the Clover would have been.— These seeds, however, were kept for 3 days in moist pellets damp with gastric juice after being ejected which would have helped to have injured them.—

Lately I have been looking during few walks at excrement of small birds; I have found 6 kinds of seeds, which is more than I expected. Lastly I have had a partride with 22 grains of dry earth on *one* foot, & to my surprise a pebble as big as a tare seed; & I now understand how this is possible for the bird scartches itself, & little plumose feathers make a sort of very tenacious plaister. Think of the millions of migratory quails, & it w^{d} be strange if some plants have not been transported across good arms of the sea. ...

Adios, my dear Hooker, I thank you most honestly for your assistance,—assistance by the way now spread over some dozen years.—

Farewell | C. Darwin ...

1857

To W. D. Fox   8 February [1857]

Down Bromley Kent
Feb. 8th

My dear Fox

I was very glad to get your note; but it was really too bad of you not to say one single word about your own health. Do you think I do not care to hear?—

It was a complete oversight that I did not write to tell you that Emma produced under blessed Chloroform our sixth Boy almost two months ago.[1] I daresay you will think only half-a-dozen Boys a mere joke; but there is a rotundity in the half-dozen which is tremendously serious to me.— Good Heavens to think of all the sendings to School & the Professions afterwards: it is dreadful. . . .

We shall be most heartily rejoiced to see you here at any time: we have now R^y to Beckenham which cuts of 2 miles & gladly will we send you both ways at any time.

But the other morning I was telling my Boys about some of our ancient entomological expeditions to Whittlesea meer &c; & how we two used to drink our tea & Coffee together daily. We had not then $20\frac{3}{4}$ children between us; & I had no stomach.

I do not think I shall have courage for Water Cure again: I am now trying mineral Acids, with, I think, good effect. I am not so well as I was a year or two ago.

I am working very hard at my Book, perhaps too hard. It will be very big & I am become most deeply interested in the way facts fall into groups. I am like Crœsus overwhelmed with my riches in facts. & I mean to make my Book as perfect as ever I can. I shall not go to press at soonest for a couple of years.

Thanks about W. Indies. I have just had a Helix pomatia withstand 14 days well in Salt-water; to my very great surprise.

I work all my friends: Are there any Mormodes at Oulton Hot-houses or any of those Orchideæ which eject their pollen-masses when irritated: if so will you examine & see what would be effect of Humble-Bee visiting flower: w^d. pollen-mass ever adhere to Bee, or w^d. it always hit direct the stigmatic surface?—

You ask about Pigeons: I keep at work & skins are now flocking in from all parts of world.—

You ask about Erasmus & my sisters: the latter have been tolerable; but Eras. not so well with more frequent fever fits & a good deal debilitated: Charlotte Langton has been very ill with Asthma & Bronchitis; but I hope is recovering.

Farewell, my dear old Friend. | Yours affect^y | C. Darwin

Are castrated Deer larger than ordinary Bucks? Do you know?

To Charles Lyell   11 February [1857]

Down Bromley Kent
Feb. 11

My dear Lyell

I was glad to see in the newspapers about the Austrian Expedition:[2] ... I do not know whether the Expedition is tied down to call at only fixed spots. But if there be any choice or power in the scientific men to influence the places, this w^d. be very desirable; it is my most delibreate conviction that nothing would aid more Natural History, than careful collecting & investigating *all the productions* of the most insulated islands, especially of the southern hemisphere.— Except Tristan d'Acunha & Kerguelen Land, they are very imperfectly known; & even at Kerguelen land, how much there is to make out about the lignite beds, & whether there are signs of old Glacial action— Every sea-shell & insects & plant is of value from such spots.

Someone in Expedition *especially* ought to have Hookers N. Zealand Essay. What grand work to explore Rodriguez with its fossil birds & little known productions of every kind—

Again the Seychelles, which with the Cocos de mar, must be a remnant of some older land.— The outer isl^d of Juan Fernandez is little known.— The investigation of these little spots by a band of naturalists would be grand.— St. Pauls & Amsterdam would be glorious botanically & geologically.— Can you not recommend them to get my Journal (& Volcanic islands) *on account of Galapagos*. If they come from North, it will be a shame & sin if they do not call at Cocos islet, N. of the Galapagos.— I always regretted that I was not able to examine the great craters on Albemarle Is^d, one of the Galapagos.

In New Zealand urge on them to look out for erratic Boulders, & marks of old Glaciers.—

Urge the use of the Dredge in Tropics; how little or nothing we know of limit of life downwards in the hot seas.—

My present work leads me to perceive how much the domestic animals have been neglected in out-of-the way countries.—

The Revilligago isl^d off Mexico, I believe, have never been trodden by foot of naturalist

If the expedition sticks to such places as Rio, C. Good Hope, Ceylon & Australia &c, it will not do much.—

Ever yours most truly | C. Darwin

I have just had Helix Pomatia quite alive & hearty after *20* days under sea-water; & this same individual about six-weeks ago had a bath of 7 days.— ...

To W. E. Darwin   [17 February 1857]

[Down]
Tuesday Night.

My dear Willy

I am very glad indeed to hear that you are in the sixth; & I do not care how difficult you find the work: am I not a kind Father? I am even almost as glad to hear of the Debating Society, for it will stir you up to read.— Do send me as soon as you can the subjects; & I will do my very best to give you hints; & mamma will try also.— But I fear, as the subjects will generally be historical or political, that I shall not be of much use.— By thinking at odds & ends of times on any subject, especially if you read a little about it, you will form some opinion & find something to say; & in truth the habit of speaking will be of greatest importance to you. Uncle Harry was here this morning, & we were telling him that we had settled for you to be a Barrister (he was one) & his first question was, "has he the gift of the gab"? But then he added, he has got industry, & that is by far the most important of all.— Mamma desires that you will read the Chapters *very well* ; & the dear old Mammy must be obeyed. Her lip is plaistered up, so we cannot tell yet how she will look. ...

Lenny, Franky & Co^y. were rather awe-struck to hear that you had bought a cane to whip the Boys.—

Be sure tell me about the Optics—& how you get on with the Reading in Chapel. Read slow & read the chapter two or three times over to yourself first; *that will make a great difference*. When I was Secretary to the Geolog. Soc, I had to read aloud to Meeting M.S. papers; but I always read them over carefully first; yet I was so nervous at first, I somehow could see nothing all around me, but the paper, & I felt as if my body was gone, & only my head left.— . . . When you write tell me how long the Boys make their speeches, & whether many get up & answer.—

Good night, my dear old fellow & future Lord Chancellor of all England.—

Your's' most affectionately | Ch. Darwin

To W. D. Fox   22 February [1857]

Down Bromley Kent
Feb. 22^{d}.

My dear Fox

I am much obliged for your various enclosures, viz (1st) about yourself, & most heartily glad I am that D^{r}. Gully has done you some good.—

Emma desires to be most kindly remembered to M^{rs}. Fox & we are very glad that she & the little girl are both well.—

I hope that your nephew may succeed in finding some lizard eggs; for it seems that he will try his best to ascertain the point in question.— By the way I have just had Helix Pomatia quite healthy after 20 days submersion in salt-water.—

Thanks about Pea case: it is a very great puzzle to me; for if I could trust to my observations on Bees, I cannot see how they can avoid being crossed; but the evidence certainly preponderates on your side, & most heavily in case of Sweet Peas.— I suppose the Queen Pea flowered at same time with adjoining Peas: are you sure of this?

With respect to Clapham School: I think favourably of it: the Boys are not so exclusively kept to Classics: arithmetic is made much of: all are taught drawing, & some modern languages.— I was rather frightened by having heard that it was rather a rough school; but young Herschel did not agree to this; & Georgy is rather a soft Boy & I cannot find out that he has anything to complain of, though of a very home-sick, disposition. I will at any time answer any queries in detail: I do not know, but could find out, whether Clergymen's sons are charged less.—

My wife agrees very heartily with your preachment against over-work, & wishes to go to Malvern; but I doubt: yet I suppose I shall take a little holiday sometime; perhaps to Tenby: though how I can leave all my experiments, I know not.—

I am got most deeply interested in my subject; though I wish I could set less value on the bauble fame, either present or posthumous, than I do, but not, I think, to any extreme degree; yet, if I know myself, I would work just as hard, though with less gusto, if I knew that my Book w^d be published for ever anonymously

Farewell, my dear Fox | Ever yours | C. Darwin

To J. D. Hooker   15 March [1857]

Down Bromley Kent

My dear Hooker

... I asked A. Gray whether he c^d. tell me about Trees in U. States; & I told him that I had expected they w^d have sexes tending to be separate from theoretical notions, & I told him result for Britain & N. Zealand from you.—

I have been thinking over your casual remarks at the Club, versus "accidental" dispersal, in contradistinction to dispersal over land *more or less* continuous; & your remarks do not quite come up to my wishes; for I want to hear whether plants offer any *positive* testimony in favour of continuous land.— Your remarks were that the dispersal & more especially *non*-dispersal could not be accounted for by "accidental" means; which of course I must agree to & can say only that we are quite ignorant of means of *trans-oceanic* transport. But then all these arguments seem to me to tell equally against "continuous more or less" land; & you must say that some were created since separation on mainlands, & some extinct since on island.— Between these excuses on both sides, there seems not much to choose, but I prefer my answer to yours.—

The same remark, seems to me applicable to your observation on the commonest species not having been transported; for it seems bold hypothesis to suppose that the commonest have been generally last created on the mainland or soonest extinguished on the island.— But I sh^d. like to hear whether you are prepared on reflexion to uphold this doctrine of the commonest being least widely disseminated on outlying isl^{ds}.— I know it holds in New Zealand & feebly owing to distance in Tristan d'Acunha., but generally I sh^d. have taken from De Candolle a different impression:— I am referring only to *identical* species in these remarks.—

What I sh^d. call positive evidence would be if proportions of Families had been exactly same on island with mainland.— If all plants were common to some mainland & island (as in your Raoul Is^d.) more especially if some *other* main-land was nearer.— If soundings concurred with any great predominance of species from any country—or any other such argument of which I know nothing.

Do not answer me, without you feel inclined, but keep this part of subject before your mind for some future essay. I have written at this length that you may see, what I for one sh^d. like to see discussed. But I will stop for I could go on prosing for another hour.— ...

So adios | Ever yours | C. Darwin

March 15th.—

To J. D. Hooker   12 April [1857]

Down Bromley Kent
Ap. 12th

My dear Hooker

Your letter has pleased me much, for I never can get it out of my head, that I take unfair advantage of your kindness, as I receive all & give nothing. What a splendid discussion you could write on whole subject of variation! The cases discussed in your last note are valuable to me, (though odious & damnable) as showing how profoundly ignorant we are on causes of variation.— ...

I have just been putting my notes together on variations *apparently* due to the immediate & direct action of external causes; & I have been struck with one result. The most firm stickers for independent creation admit, that the fur of *same* species is thinner towards south of range of same species than to north—that *same* shells are brighter coloured to S. than N.; that same is paler-coloured in deep water—that insects are smaller & darker on mountains—more lurid & testaceous near sea—that plants are smaller & more hairy & with brighter flowers on mountains: now in all such (& other cases) cases, distinct species in the two zones follow the same rule, which seems to me to be most simply explained by species, being only strongly marked varieties, & therefore following same laws as recognised & admitted varieties. I mention all this on account of variation of plants in ascending mountains; I have quoted the foregoing remark only generally with no examples, for I add there is so much doubt & dispute what to call varieties; but yet I have stumbled on so many casual remarks on *varieties* of plants on mountains being so characterised, that I presume there is some truth in it. What think you? do you believe there is *any* tendency in *varieties*, as *generally* so called, of plants to become more hairy & with proportionally larger & brighter coloured flowers in ascending a mountain.—

I have been interested in my "weed garden" of 32 feet square: I mark each seedling as it appears, & I am astonished at number that come up. & still more at number killed by slugs &c.— Already 59 have been so killed; I expected a good many, but I had fancied that this was a less potent check than it seems to be; & I attributed almost exclusively to mere choking the destruction of seedlings.— Grass-seedlings seem to suffer much less than exogens.—

I have *almost* finished my floating experiments on salt-water: $\frac{72}{94}$ sunk under 10 days—seven plants, however, floated *on average* 67 days each. ...

I *think* it will turn out on average from my very few experiments, of **very** little value, but better than mere conjecture, that about 110 of all plants of a country will float when dryed 30 days *& the seeds then germinate*; & this on *average* current of 33 miles per day will carry them a good way. I would wager that the pods of the Acacia(?) scandens which get to the Azores had been dried first.— I suppose the oriental species does not fruit at Kew: if it did, I sh^d. like to try. ...

Farewell | C. Darwin

P.S Strictly according to my experiments a little above $\frac{1}{7}$ (.140) of the plants of any country could be transported 924 miles & *would then germinate!* for $\frac{18}{94}$ have floated above 28 days & $\frac{64}{87}$ is proportion of seeds which germinate after 28 days immersion.— & average of current in Atlantic is 33 miles per diem.— ...

To Charles Lyell   13 April [1857]

Down Bromley Kent
Ap. 13th

My dear Lyell

I have been particularly glad to see Wollaston's letter. The news did not require any breaking to me; for though as a general rule I am much opposed to the Forbesian continental extensions, I have no objection whatever to its being proved in some cases. Not that I can admit that W. has by any means proved it; nor, I think, can anyone else, till we know something of the means of distribution of insects.— But the close similarity or identity of the two Faunas is certainly very interesting.— I am extremely glad to hear that your Madeira paper is making progress; & I shall be most curious to see. I sh^d. be infinitely obliged for a separate copy, whenever printed.— My health has been very poor of late, & I am going in a week's time for a fortnight of hydropathy & rest.— My everlasting species-Book quite overwhelms me with work— It is beyond my powers, but I hope to live to finish it.—

Farewell | My dear Lyell | Ever yours | C. Darwin

To Philip Henry Gosse   27 April [1857]

Moor Park | Farnham | Surrey

April 27th

My dear Sir

I have thought that perhaps in course of summer you would have an opportunity & would be so very kind as to try a *little* experiment for me.— I think I can tell best what I want, by telling what I have done. The wide distribution of same species of F. Water Molluscs has long been a great perplexity to me: I have just lately hatched a lot & it occurred to me that when first born they might perhaps have not acquired phytophagous habits, & might perhaps like nibbling at a Ducks-foot.— Whether this is so I do not know, & indeed do not believe it is so, but I found when there were many *very* young Molluscs in a *small* vessel with aquatic plants, amongst which I placed a dried Ducks foot, that the little barely visible shells **often** crawled over it, & that they *adhered* so firmly that they c^d. not be shaken off, & that the foot being kept out of water in a damp atmosphere, the little Molluscs survived well 10, 12 & 15 hours & a *few* even 24 hours.— And thus, I believe, it must be that Fr. W. shells get from pond to pond & even to islands out at sea. A Heron fishing for instance, & then startled might well on a rainy day carry a young mollusc for a long distance.—

Now you will remember that E. Forbes argues chiefly from the difficulty of imagining how *littoral* sea-molluscs could cross tracts of open ocean, that islands, such as Madeira must have been joined by continuous land to Europe: which seems to me, for many reasons, very rash reasoning.— Now what I want to beg of you, is, that you would try an analogous experiment with some sea-molluscs, especially any strictly littoral species,—hatching them in numbers in a smallish vessel & seeing whether, either *in larval* or *young shell state* they can *adhere* to a birds foot & survive say 10 hours in *damp* atmosphere out of water. It may seem a trifling experiment, but seeing what enormous conclusions poor Forbes drew from his belief that he knew all means of distribution of sea-animals, it seems to me worth trying.—

My health has lately been very indifferent, & I have come here for a fortnight's water-cure.— ...

I hope you will forgive my troubling you on the above point & believe me, | My dear Sir | Your's very sincerely | Ch. Darwin

P.S. | Can you tell me, you who have so watched all sea-creatures, whether male Crustaceans ever fight for the females: is the female sex in the sea, like on the land, "teterrima belli causa"?[3]

I beg you not to answer this letter, without you can & will be so kind as to tell about Crustacean Battles, if such there be.—

To Alfred Russel Wallace   1 May 1857

Down Bromley Kent [Moor Park, Surrey]

May 1.— 1857

My dear Sir

I am much obliged for your letter of Oct. 10th. from Celebes received a few days ago: in a laborious undertaking sympathy is a valuable & real encouragement. By your letter & even still more by your paper in Annals, a year or more ago,[4] I can plainly see that we have thought much alike & to a certain extent have come to similar conclusions. In regard to the Paper in Annals, I agree to the truth of almost every word of your paper; & I daresay that you will agree with me that it is very rare to find oneself agreeing pretty closely with any theoretical paper; for it is lamentable how each man draws his own different conclusions from the very same fact.—

This summer will make the 20th year (!) since I opened my first-note-book, on the question how & in what way do species & varieties differ from each other.— I am now preparing my work for publication, but I find the subject so very large, that though I have written many chapters, I do not suppose I shall go to press for two years.—

I have never heard how long you intend staying in the Malay archipelago; I wish I might profit by the publication of your Travels there before my work appears, for no doubt you will reap a large harvest of facts.— I have acted already in accordance with your advice of keeping domestic varieties & those appearing in a state of nature, distinct; but I have sometimes doubted of the wisdom

of this, & therefore I am glad to be backed by your opinion.—
I must confess, however, I rather doubt the truth of the now very
prevalent doctrine of all our domestic animals having descended from
several wild stocks; though I do not doubt that it is so in some
cases.— I think there is rather better evidence on the sterility of
Hybrid animals that you seem to admit: & in regard to Plants the
collection of carefully recorded facts by Kölreuter & Gærtner, (&
Herbert) is *enormous*.—

I most entirely agree with you on the little effects of "climatal
conditions", which one sees referred to ad nauseam in all Books; I
suppose some very little effect must be attributed to such influences,
but I fully believe that they are very slight.— It is really *impossible* to
explain my views in the compass of a letter on the causes & means
of variation in a state of nature; but I have slowly adopted a distinct
& tangible idea.— Whether true or false others must judge; for the
firmest conviction of the truth of a doctrine by its author, seems, alas,
not to be slightest guarantee of truth.—

I have been rather disappointed at my results in the Poultry line;
but if you sh^d. after receiving this stumble on any curious domestic
breed, I sh^d be very glad to have it; but I can plainly see that this
result will not be at all worth the trouble which I have taken.—
The case is different with the domestic Pigeons; from its study I have
learned much.— The Rajah has sent me some of his Pigeons & Fowls
& **Cats** skins from interior of Borneo, & from Singapore.—

Can you tell me positively that Black Jaguars or Leopards are
believed generally or always to pair with Black? I do not think colour
of offspring good evidence.— Is the case of parrots fed on fat of
fish turning colour, mentioned in your Travels? I remember case of
Parrot with, (*I think*,) poison from some Toad put into hollow whence
primaries had been removed.

One of the subjects on which I have been experimentising & which
cost me much trouble, is the means of distribution of all organic
beings found on oceanic islands; & any facts on this subject would
be most gratefully received: Land-Molluscs are a great perplexity to
me.—

This is a very dull letter, but I am a good deal out of health; &
am writing this, not from my home, as dated, but from a water-cure
establishment.

With most sincere good wishes for your success in every way I
remain | My dear Sir | Yours sincerely | Ch. Darwin

To J. D. Hooker [2 May 1857]

Moor Park
Saturday

My dear Hooker

... How candidly & meekly you took my Jeremiad on your severity to second class men. After I had sent it off, an ugly little voice asked me once or twice how much of my noble defence of the poor in spirit & in fact, was owing to your having not seldom smashed favourite notions of my own.— I silenced the ugly little voice with contempt, but it would whisper again & again.— I sometimes despise myself as a poor compiler, as heartily as you could do, though I do *not* despise my whole work, as I think there is enough known to lay a foundation for the discussion on origin of species.— I have been led to despise & laugh at myself as compiler, for having put down that "alpine plants have large flowers," & now perhaps I may write over these very words "alpine plants have small or apelatous flowers"! ...

With thanks for your never failing assistance to me— Ever yours | My dear Hooker | C. Darwin

I return home, thanks be to God, on Wednesday.—

I remember that you were surprised at number of seeds germinating in pond mud: I tried a 4th. Pond, & took about as much mud, (rather more than in former cases) as would fill a very large breakfast cup, & before I had left home 118 plants had come up; how many more will be up on my return I know not.— This bears on chance of Birds by their muddy feet transporting F.W. plants.—

It w^{d}. not be a bad dodge for a collector in country, when plants were *not* in seed, to collect & dry mud from Ponds.

To J. D. Hooker 3 June [1857]

Down Bromley Kent
June 3^{d}

My dear Hooker

I am going to enjoy myself by having a prose on my own subjects to you, & this is a greater enjoyment to me than you will readily understand; as I for months together do not open my mouth on Nat. History. ...

My observations, though on so infinitely a small scale, on the struggle for existence, begin to make me see a little clearer how the fight goes on: out of 16 kinds of seed sown on my meadow, 15 have germinated, but now they are perishing at such a rate that I doubt whether more than one will flower. Here we have choking, which has taken place likewise on great scale with plant not seedlings in a bit of my lawn allowed to grow up. On other hand in a bit of ground 23 feet, I have daily marked each seedling weed as it has appeared during March, April & May, and 357 have come up, & of these 277 have *already* been killed chiefly by slugs.— By the way at Moor Park, I saw rather pretty case of effect of animals on vegetation: there are enormous commons with clumps of old Scotch firs on hills, & about 8–10 years ago some of these commons were enclosed & all round the clumps nice young trees are springing up by the millions, looking exactly as if planted so many are of same age. In other part of common, not yet enclosed, I looked for miles & not *one* young tree c^d be seen; I then went near (within $\frac{1}{4}$ of mile of the clumps & looked closely in the heather, & there I found tens of thousands of young scotch-firs (30 in one square yard) with their tops nibbled off by the few cattle which occasionally roam over these wretched Heaths. One little tree 3 inches high, by the rings appeared to be 26 years old with a short stem about as thick as stick of sealing wax.— What a wondrous problem it is,—what a play of forces, determining the kinds & proportions of each plant in a square yard of turf! It is to my mind truly wonderful. And yet we are pleased to wonder when some animal or plant becomes extinct. . . .

Good Bye | My dear Hooker | Ever yours | Ch. Darwin . . .

I believe you are afraid to send me a ripe Edwardsia pod for fear I sh^d float it from N. Zealand to Chile!!!

To J. D. Hooker   5 June [1857]

Down Bromley Kent
June 5th

My dear Hooker

. . . It is an old notion of mine that more good is done by giving medals to younger men in the early part of career, than as a mere reward to men whose scientific career is nearly finished.— Whether medals ever do any good is question which does not concern us, as there the medals are.— I am almost inclined to think I would rather lower standard & give medal to young workers than to old ones, with no *especial* claims. With regard to *especial* claims, I think it just deserves your attention, that if general claims are once admitted, it opens the door to great laxity in giving them.— Think of case of very rich man who aided *solely* with his money but to a grand extent—or to such an inconceivable prodigy as a minister of Crown who really cared for science. Would you give such men medals—perhaps medals could not be better applied than *exclusively* to such men! I confess at present I incline to stick to especial claims, which can be put down on paper. . . .

I have been so much interested this morning in comparing all my notes on the variation of the several species of genus Equus & the results of their crossing: Taking most strictly analogous facts amongst the blessed Pigeons for my guide, I believe I can plainly see the colouring & marks of the grandfather of the Ass, Horse, Quagga, Hemionus & Zebra, some millions of generations ago! Should not I sneer at any one who made such a remark to me a few years ago!—but my evidence seems to me so good, that I shall publish my vision at end of my little discussion on this genus.

I have of late inundated you with my notions, you best of friends & philosophers.—

Adios | C. Darwin . . .

To John Lubbock   14 [July 1857]

Down.—

14th

My dear Lubbock

You have done me the greatest possible service in helping me to clarify my Brains. If I am as muzzy on all subjects as I am on proportions & chance,—what a Book I shall produce!—

I have divided N. Zealand Flora as you suggested. There are 339 species in genera of 4 & upwards [species] & 323 in genera of 3 & less. The 339 species have 51 species presenting one or more varieties— The 323 species have only 37: proportionally (as 339:323 :: 51.:48.5) they ought to have had 48$\frac{1}{2}$ species presenting vars.— So that the case goes as I want it, but not strong enough, without it be general, for me to have much confidence in.

I am quite convinced yours is the right way; I had thought of it, but sh^{d} never have done it, had it not been for my most fortunate conversation with you.

I am quite shocked to find how easily I am muddled, for I had before thought over the subject much, & concluded my way was fair. It is dreadfully erroneous. What a disgraceful blunder you have saved me from. I heartily thank you—

Ever yours | C. Darwin

It is enough to make me tear up all my M.S. & give up in despair.—

It will take me several weeks to go over all my materials. But oh if you knew how thankful I am to you.—

To Asa Gray   5 September [1857]

Down Bromley Kent
Sept. 5th

My dear Gray

I forget the exact words which I used in my former letter, but I daresay I said that I thought you would utterly despise me, when I told you what views I had arrived at, which I did because I thought I was bound as an honest man to do so. . . .

I did not feel in the least sure that when you knew whither I was tending, that you might not think me so wild & foolish in my views (God knows arrived at slowly enough, & I hope conscientiously) that you would think me worth no more notice or assistance. To give one example, the last time I saw my dear old friend Falconer, he attacked me most vigorously, but quite kindly, & told me "you will do more harm than any ten naturalists will do good"— "I can see that you have already *corrupted* & half-spoiled Hooker"(!!). Now when I see such strong feeling in my oldest friends, you need not wonder that I always expect my views to be received with contempt. But enough & too much of this.— . . .

As you seem interested in subject, & as it is an *immense* advantage to me to write to you & to hear **ever so briefly**, what you think, I will enclose (*copied* so as to save you trouble in reading) the briefest abstract of my notions on the **means** by which nature makes her species. Why I think that species have really changed depends on general facts in the affinities, embryology, rudimentary organs, geological history & geographical distribution of organic beings. In regard to my abstract you must take immensely on trust; each paragraph occupying one or two chapters in my Book. You will, perhaps, think it paltry in me, when I ask you not to mention my doctrine; the reason is, if anyone, like the Author of the Vestiges, were to hear of them, he might easily work them in, & then I sh^d have to quote from a work perhaps despised by naturalists & this would greatly injure any chance of my views being received by those alone whose opinion I value.— . . .

My dear D^r. Gray | Believe me with much | sincerity Your's truly | C. Darwin . . .

[Enclosure]

I. It is wonderful what the principle of Selection by Man, that is the picking out of individuals with any desired quality, and breeding from them, and again picking out, can do. Even Breeders have been astonished at their own results. They can act on differences inappreciable to an uneducated eye. Selection has been *methodically* followed in *Europe* for only the last half century. But it has occasionally, and even in some degree methodically, been followed in the most ancient times. There must have been, also, a kind of unconscious selection from the most ancient times,—namely in the preservation of the individual animals (without any thought of their offspring) most useful to each race of man in his particular circumstances. The "rogueing" as nurserymen call the destroying of varieties, which depart from their type, is a kind of selection. I am convinced that intentional and occasional selection has been the main agent in making our domestic races. But, however, this may be, its great power of modification has been indisputably shown in late times. Selection acts only by the accumulation of very slight or greater variations, caused by external conditions, or by the mere fact that in generation the child is not absolutely similar to its parent. Man by this power of accumulating variations adapts living beings to his wants,—he *may be said* to make the wool of one sheep good for carpets and another for cloth &c.—

II. Now suppose there was a being, who did not judge by mere external appearance, but could study the whole internal organization—who never was capricious,—who should go on selecting for one end during millions of generations, who will say what he might not effect! In nature we have some *slight* variations, occasionally in all parts: and I think it can be shown that a change in the conditions of existence is the main cause of the child not exactly resembling its parents; and in nature geology shows us what changes have taken place, and are taking place. We have almost unlimited time: no one but a practical geologist can fully appreciate this: think of the Glacial period, during the whole of which the same species of shells at least have existed; there must have been during this period millions on millions of generations.

III. I think it can be shown that there is such an unerring power at work, or *Natural Selection* (the title of my Book), which selects exclusively for the good of each organic being. The elder De Candolle, W. Herbert, and Lyell have written strongly on the struggle for life; but even they have not written strongly enough. Reflect that every being (even the Elephant) breeds at such a rate, that in a few years, or at most a few centuries or thousands of years, the surface of the earth would not hold the progeny of any one species. I have found it hard constantly to bear in mind that the increase of every single species is checked during some part of its life, or during some shortly recurrent generation. Only a few of those annually born can live to propagate their kind. What a trifling difference must often determine which shall survive and which perish—

IV. Now take the case of a country undergoing some change; this will tend to cause some of its inhabitants to vary slightly; not but what I believe most beings vary at all times enough for selection to act on. Some of its inhabitants will be exterminated, and the remainder will be exposed to the mutual action of a different set of inhabitants, which I believe to be more important to the life of each being than mere climate. Considering the infinitely various ways, beings have to obtain food by struggling with other beings, to escape danger at various times of life, to have their eggs or seeds disseminated &c. &c, I cannot doubt that during millions of generations individuals of a species will be born with some slight variation profitable to some part of its economy; such will have a better chance of surviving, propagating, this variation, which again will be slowly increased by the accumulative action of Natural selection; and the variety thus formed will either coexist with, or more commonly will exterminate its parent form. An organic being like the woodpecker or misletoe may thus come to be adapted to a score of contingencies: natural selection, accumulating those slight variations in all parts of its structure which are in any way useful to it, during any part of its life.

V. Multiform difficulties will occur to everyone on this theory. Most can I think be satisfactorily answered.— "Natura non facit saltum" answers some of the most obvious.— The slowness of the change, and only a very few undergoing change at any one time answers others. The extreme imperfections of our geological records answers others.—

VI. One other principle, which may be called the principle of divergence plays, I believe, an important part in the origin of species. The same spot will support more life if occupied by very diverse forms: we see this in the many generic forms in a square yard of turf (I have counted 20 species belonging to 18 genera),—or in the plants and insects, on any little uniform islet, belonging almost to as many genera and families as to species.— We can understand this with the higher, animals whose habits we best understand. We know that it has been experimentally shown that a plot of land will yield a greater weight, if cropped with several species of grasses than with 2 or 3 species. Now every single organic being, by propagating so rapidly, may be said to be striving its utmost to increase in numbers. So it will be with the offspring of any species after it has broken into varieties or sub-species or true species. And it follows, I think, from the foregoing facts that the varying offspring of each species will try (only few will succeed) to seize on as many and as diverse places in the economy of nature, as possible. Each new variety or species, when formed will generally take the places of and so exterminate its less well-fitted parent. This, I believe, to be the origin of the classification or arrangement of all organic beings at all times. These always **seem** to branch and sub-branch like a tree from a common trunk; the flourishing twigs destroying the less vigorous,—the dead and lost branches rudely representing extinct genera and families.

This sketch is *most* imperfect; but in so short a space I cannot make it better. Your imagination must fill up many wide blanks.— Without some reflexion it will appear all rubbish; perhaps it will appear so after reflexion.— | C. D.

This little abstract touches only on the accumulative power of natural selection, which I look at as by far the most important element in the production of new forms. The laws governing the incipient or primordial variation (unimportant except as to groundwork for selection to act on, in which respect it is all important) I shall discuss under several heads, but I can come, as you may well believe, only to very partial & imperfect conclusions.—

To T. H. Huxley   26 September [1857]

Down Bromley Kent
Sept. 26^(th)

My dear Huxley

Thanks for your very pleasant note.— It amuses me to see what a bug-bear I have made myself to you; when having written some very pungent & good sentences it must be very disagreeable to have my face rise up like an ugly ghost.— I have always suspected Agassiz of superficiality & wretched reasoning powers; but I think such men do immense good in their way. See how he stirred up all Europe about Glaciers.— By the way Lyell has been at the Glaciers, or rather their effects, & seems to have done good work in testing & judging what others have done. . . .

Farewell | Yours very truly | C. Darwin

In regard to Classification, & all the endless disputes about the "Natural System which no two authors define in same way, I believe it ought, in accordance to my heteredox notions, to be simply genealogical.— But as we have no written pedigrees, you will, perhaps, say this will not help much; but I think it ultimately will, whenever heteredoxy becomes orthodoxy, for it will clear away an immense amount of rubbish about the value of characters &—will make the difference between analogy & homology, clear.— The time will come I believe, though I shall not live to see it, when we shall have very fairly true genealogical trees of each great kingdom of nature. . . .

To T. H. Huxley   3 October [1857]

Down Bromley Kent
Oct. 3^{d}.

My dear Huxley.

I know you have no time for speculative correspondence; & I did not **in the least** expect an answer to my last. But I am very glad to have had it, for in my eclectic work, the opinions of the few good men are of great value to me.—

I knew, of course, of the Cuvierian view of Classification, but I think that most naturalists look for something further, & search for "the natural system",—"for the plan on which the Creator has worked" &c &c.— It is this further element which I believe to be simply genealogical.

But I sh^{d}. be very glad to have your answer (either *when we meet* or by note) to the following case, *taken by itself* & *not allowing yourself to look any further than to the point in question.*

Grant all races of man descended from one race; *grant* that all structure of each race of man were perfectly known—**grant** that a perfect table of descent of each race was perfectly known.— grant all this, & then do you not think that most would prefer as the best classification, a genealogical one, even if it did occasionally put one race not quite so near to another, as it would have stood, if allocated by structure alone. Generally, we may safely presume, that the resemblance of races & their pedigrees would go together.

I sh^{d}. like to hear what you w^{d}. say on this purely theoretical case.

Ever your's very truly | C. Darwin

It might be asked why is development so all-potent in classification, as I fully admit it is: I believe it is, because it depends on, & best betrays, genealogical descent; but this is too large a point to enter on.

To Asa Gray   29 November [1857]

Down Bromley Kent
Nov. 29th

My dear Gray

This shall be such an extraordinary note as you have never received from me, for it shall not contain *one* single question or request. I thank you for your impression on my views. Every criticism from a good man is of value to me. What you hint at generally is very very true, that my work will be grievously hypothetical & large parts by no means worthy of being called inductive; my commonest error being probably induction from too few facts.— I had not thought of your objection of my using the term "natural Selection" as an agent; I use it much as a geologist does the word Denudation, for an agent, expressing the result of several combined actions. I will take care to explain, not merely by inference, what I mean by the term; for I must use it, otherwise I sh^{d.} incessantly have to expand it into *some such* (here miserably expressed) formula as the following, "the tendency to the preservation (owing to the severe struggle for life to which all organic beings at some time or generation are exposed) of any the slightest variation in any part, which is of the slightest use or favourable to the life of the individual which has thus varied; together with the tendency to its inheritance". Any variation, which was of no use whatever to the individual, would not be preserved by this process of "natural selection". But I will not weary you by going on; as I do not suppose I c^d make my meaning clearer without large expansion.— I will only add one other sentence: several varieties of Sheep have been turned out together on the Cumberland Mountains, & one particular breed is found to succeed so much better than all the others, that it fairly starves the others to death: I sh^{d.} here say that natural selection picks out this breed, & would *tend* to improve it or aboriginally to have formed it. ...

By the way I must tell you what I heard yesterday, though not in your line, but on subject of the crossing of individuals. Barnacles (Balanus) are hermaphrodite & with their well shut up shell offer as great a difficulty to crossing *as can well be conceived*: I found an individual with monstrous & *imperforate* penis, but yet with fertilised ova; but I did not know whether it might not be case of parthogenesis or a strange accident of some floating spermatozoa; well yesterday I had an account by a man who watching some shells, saw one protrude its long prosciformed penis, & insert it in the shell of an adjoining individual! So here is a load off my mind.—

You speak of species not having any material base to rest on; but is this any greater hardship than deciding what deserves to be called a variety & be designated by a greek letter. When I was at systematic work, I know I longed to have no other difficulty (great enough) than deciding whether the form was distinct enough to deserve a name; & not to be haunted with undefined & unanswerable question whether it was a true species. What a jump it is from a well marked variety, produced by natural cause, to a species produced by the separate act of the Hand of God. But I am running on foolishly.— By the way I met the other day Phillips, the Palæontologist, & he asked me "how do you define a species?"— I answered "I cannot" Whereupon he said "at last I have found out the only true definition,—'any form which has ever had a specific name"! . . .

But I am amusing myself by scribbling away all my notions without any mercy.

Forgive me, & believe | My dear Gray | Yours heartily obliged | C. Darwin

How I wish I knew what large, (for large it must be) Moth or Humble Bee visits & fertilises Lobelia fulgens in its native home: do you know any southern young Botanist who w^d look to this? I would cover a plant with a very coarse gauze cap, & then not a pod would set I believe. But by Jove I have broken my vow by a sort of question or request!

To A. R. Wallace   22 December 1857

Down Bromley Kent.

Dec. 22/57

My dear Sir

I thank you for your letter of Sept. 27th.— I am extremely glad to hear that you are attending to distribution in accordance with theoretical ideas. I am a firm believer, that without speculation there is no good & original observation. Few travellers have ⟨at⟩tended to such points as you are now at work on; & indeed the whole subject of distribution of animals is dreadfully behind that of Plants.— You

183

say that you have been somewhat surprised at no notice having been taken of your paper in the Annals: I cannot say that I am; for so very few naturalists care for anything beyond the mere description of species. But you must not suppose that your paper has not been attended to: two very good men, Sir C. Lyell & Mr E. Blyth at Calcutta specially called my attention to it. Though agreeing with you on your conclusion⟨s⟩ in that paper, I believe I go much further than you; but it is too long a subject to enter on my speculative notions.—

I have not yet seen your paper on distribution of animals in the Arru Isl^{ds}:— I shall read it with the **utmost** interest; for I think that the most interesting quarter of the whole globe in respect to distribution; & I have long been very imperfectly trying to collect data for the Malay archipelago.—

I shall be quite prepared to subscribe to your doctrine of subsidence: indeed from the quite independent evidence of the Coral Reefs I coloured my original map in my Coral volume of the Arru Isl^d. as one of subsidence, but got frightened & left it uncoloured.— But I can see that you are inclined to go **much** further than I am in regard to the former connections of oceanic islands with continent: Ever since poor E. Forbes propounded this doctrine, it has been eagerly followed; & Hooker elaborately discusses the former connections of all the Antarctic isl^{ds} & New Zealand & S. America.— About a year ago I discussed this subject much with Lyell & Hooker (for I shall have to treat of it) & wrote out my arguments in opposition; but you will be glad to hear that neither Lyell or Hook⟨er⟩ thought much of my arguments: nevertheless for once in my life I dare withstand the almost preternatural sagacity of Lyell.—

You ask about Land-shells on islands far distant from continents: Madeira has a few identical with those of Europe, & here the evidence is really good as some of them are sub-fossil. In the Pacific isl^{ds} there are cases, of identity, which I cannot at present persuade myself to account for by introduction through man's agency; although D^r. Aug. Gould has conclusively shown that many land-shells have there been distributed over the Pacific by man's agency. These cases of introduction are most plaguing. Have you not found it so, in the Malay archipelago? it has seemed to me in the lists of mammals of Timor & other islands, that *several* in all probability have been naturalised.

Since writing before, I have experimentised a little on some land-mollusca & have found sea-water not quite so deadly as I anticipated.

You ask whether I shall discuss "man";— I think I shall avoid whole subject, as so surrounded with prejudices, though I fully admit that it is the highest & most interesting problem for the naturalist.— My work, on which I have now been at work more or less for 20 years, will *not* fix or settle anything; but I hope it will aid by giving a large collection of facts with one definite end: I get on very slowly, partly from ill-health, partly from being a very slow worker.— I have got about half written; but I do not suppose I shall publish under a couple of years. I have now been three whole months on one chapter on Hybridism!

I am astonished to see that you expect to remain out 3 or 4 years more: what a wonderful deal you will have seen; & what interesting areas,—the grand Malay Archipelago & the richest parts of S. America!— I infinitely admire & honour your zeal & courage in the good cause of Natural Science; & you have my very sincere & cordial good wishes for success of all kinds; & may all your theories succeed, except that on oceanic islands, on which subject I will do battle to the death

Pray believe me. | My dear Sir | Yours very sincerely | C. Darwin

1858

To J. D. Hooker   12 January [1858]

Down Bromley Kent
Jan. 12th

My dear Hooker

I want to ask a question which will take you only few words to answer.— It bears on my former belief (& Asa Gray strongly expressed opinion) that Pipilionaceous flowers were fatal to my notion of there being no eternal hermaphrodites.— First let me say how evidence goes: you will remember my facts going to show that Kidney Beans require visits of Bees to be fertilised. This has been positively stated to be case with Lathyrus grandiflorus, & has been very partially verified by me.— Sir W. Macarthur tells me that Erythrina will hardly seed in Australia without petals are moved as if by Bee.— I have just met statement that with common Bean, when the Humble-bees bite holes at base of flower & therefore cease visiting mouth of corolla "hardly a bean will set".

But now comes a much more curious statement that 1842–43 "since Bees were established at Wellington (N. Zealand), Clover seeds all over the Settlement, *which it did not before*". The writer evidently has no idea what the connexion can be.— Now I cannot help at once connecting this statement (& all the foregoing statements in some degree support each other, as all have been advanced without any sort of theory) with the remarkable absence of Papilionaceous plants in N. Zealand.—

I see in your list *Clianthus, Carmichælia* 4 species— A new genus, a shrub; & Edwardsia—(is latter Papilionaceous?) Now what I want to know, is, whether any of these have flowers as small as clover; for if they have large flowers they may be visited by Humble-Bee, which I think I remember do exist in N. Zealand; & which Humble Bees would not visit the smaller clovers.— Even the very minute little yellow Clover in England has every flower visited & revisited by Hive Bees, as I know by experience.— Would it not be a curious case of correlation if it could be shown to be probable that herbaceous & small Leguminosæ do not exist because **when seeds washed ashore!!!** no small Bees exist there. Though this latter fact must be ascertained! I may not prove anything, but does it not seem odd that so many quite independent facts, or rather statements should point all in one direction viz that Bees are necessary to the fertilisation of Papilionaceous flowers.

Ever yours | C. Darwin ...

To W. E. Darwin   27 [February 1858]

Down.—
27th

My dear William

The more I think of it, the more clear I am that you had better go to Christ. Coll. so I will write & enter you tomorrow. You are in error about knowing all the men of your college.— I do not think I knew even to bow to 15 men in college & was intimate with only 2 or 3 men.— Most of my friends belonged to Trinity & St. Johns & Emanuel.— I think there would be much more temptation in many ways to be idle at Trinity; & it is hard enough for the young to be industrious. You must see that when my fortune is divided amongst 8 of you, there cannot be enough for each to live comfortably & keep house, & those that do not work must be poor (though thank God with food enough) all their lives. You may rely on it, habits of industry at the University will make all the difference in your success in after life. ...

I am not very brisk, so no more my dear old man | Your affect. Father | C. Darwin

To J. D. Hooker   8 [June 1858]

Down
8th

My dear Hooker

I am confined to sofa with Boil, so you must let me write in pencil— You would laugh, if you could know how much your note pleased me. I had *firmest* conviction that you would say all my M.S was bosh,[1] & thank God you are one of the few men who dare speak truth. Though I sh^{d}. not have much cared about throwing away what you have seen, yet I have been forced to confess to myself that all was much alike, & if you condemned that you w^d. condemn all—my life's work—& that I confess made me a little low—but I c^d. have borne it, for I have the conviction that I have honestly done my best.— The discussion comes in at end of long chapter on variation in a state of nature, so that I have discussed, as far as able, what to call varieties.— I will try to leave out all allusion to genera coming in & out in this part, till when I discuss the "principle of Divergence", which with "Natural Selection" is the key-stone of my Book & I have very great confidence it is sound. I w^d. have this discussion copied out, if I could really think it would not bore you to read—for believe me I value to the full every word of criticism from you, & the advantage, which I have derived from you, cannot be told. . . .

Farewell— your Note has relieved me **immensely** | Yours ever | C. Darwin . . .

To Charles Lyell   18 [June 1858]

Down Bromley Kent
18th.

My dear Lyell

Some year or so ago, you recommended me to read a paper by Wallace in the Annals, which had interested you & as I was writing to him, I knew this would please him much, so I told him. He has to day sent me the enclosed & asked me to forward it to you.^{2} It seems to me well worth reading. Your words have come true with a vengeance that I sh^d. be forestalled. You said this when I explained to you here very briefly my views of "Natural Selection" depending on the Struggle for existence.— I never saw a more striking coincidence. if Wallace had my M.S. sketch written out in 1842 he could not have made a better short abstract! Even his terms now stand as Heads of my Chapters.

Please return me the M.S. which he does not say he wishes me to publish; but I shall of course at once write & offer to send to any Journal. So all my originality, whatever it may amount to, will be smashed. Though my Book, if it will ever have any value, will not be deteriorated; as all the labour consists in the application of the theory.

I hope you will approve of Wallace's sketch, that I may tell him what you say.

My dear Lyell | Yours most truly | C. Darwin

To Charles Lyell [25 June 1858]

Down Bromley Kent
Friday

My dear Lyell

I am very very sorry to trouble you, busy as you are, in so merely personal an affair. But if you will give me your deliberate opinion, you will do me as great a service, as ever man did, for I have entire confidence in your judgment & honour.— . . .

There is nothing in Wallace's sketch which is not written out much fuller in my sketch copied in 1844, & read by Hooker some dozen years ago. About a year ago I sent a short sketch of which I have copy of my views (owing to correspondence on several points) to Asa Gray, so that I could most truly say & prove that I take nothing from Wallace. I sh^d be *extremely* glad **now** to publish a sketch of my general views in about a dozen pages or so. But I cannot persuade myself that I can do so honourably. Wallace says nothing about publication, & I enclose his letter.— But as I had not intended to publish any sketch, can I do so honourably because Wallace has sent me an outline of his doctrine?— I would far rather burn my whole book than that he or any man sh^d think that I had behaved in a paltry spirit. Do you not think his having sent me this sketch ties my hands? I do not in least believe that that he originated his views from anything which I wrote to him.

If I could honourably publish I would state that I was induced now to publish a sketch (& I sh^d be very glad to be permitted to say to follow your advice long ago given) from Wallace having sent me an outline of my general conclusions.— We differ only, that I was led to my views from what artificial selection has done for domestic animals. I could send Wallace a copy of my letter to Asa Gray to show him that I had not stolen his doctrine. But I cannot tell whether to publish now would not be base & paltry: this was my first impression, & I sh^d have certainly acted on it, had it not been for your letter.—

This is a trumpery affair to trouble you with; but you cannot tell how much obliged I sh^d be for your advice.—

By the way would you object to send this & your answer to Hooker to be forwarded to me, for then I shall have the opinion of my two best & kindest friends.— This letter is miserably written & I write it now, that I may for time banish whole subject. And I am worn out with musing.

I fear we have case of scarlet-fever in House with Baby.—[3] Etty[4] is weak but is recovering.—

My good dear friend forgive me.— This is a trumpery letter influenced by trumpery feelings.

Yours most truly | C. Darwin

I will never trouble you or Hooker on this subject again.—

To J. D. Hooker [29 June 1858]

Down
Tuesday

My dearest Hooker

You will, & so will Mrs Hooker, be most sorry for us when you hear that poor Baby died yesterday evening. I hope to God he did not suffer so much as he appeared. He became quite suddenly worse. It was Scarlet-Fever. It was the most blessed relief to see his poor little innocent face resume its sweet expression in the sleep of death.— Thank God he will never suffer more in this world.

I have received your letters. I cannot think now on subject, but soon will. But I can see that you have acted with more kindness & so has Lyell even than I could have expected from you both most kind as you are.[5]

I can easily get my letter to Asa Gray copied, but it is too short.—

Poor Emma behaved nobly & how she stood it all I cannot conceive. It was wonderful relief, when she could let her feelings break forth—

God Bless you.— You shall hear soon as soon as I can think

Yours affectionately | C. Darwin

To J. D. Hooker [29 June 1858]

[Down]
Tuesday Night

My dear Hooker

I have just read your letter, & see you want papers at once. I am quite prostrated & can do nothing but I send Wallace & my abstract of abstract of letter to Asa Gray, which gives most imperfectly **only** *the means of change & does not touch* on reasons for believing species do change. I daresay all is too late. I hardly care about it.—

But you are too generous to sacrifice so much time & kindness.— It is most generous, most kind. I send sketch of 1844 **solely** that you may see by your own handwriting that you did read it.—

I really cannot bear to look at it.— Do not waste much time. It is miserable in me to care at all about priority.—

The table of contents will show what it is. I would make a similar, but shorter & more accurate sketch for Linnean Journal.— I will do anything

God Bless you my dear kind friend. I can write no more. I send this by servant to Kew.

Yours | C. Darwin

To Asa Gray   4 July 1858

Down Bromley Kent
July 4th.— 1858

My dear Gray

I have not answered your note of May 21 for I have had death & illness & misery amongst my children. And we are all going immediately from home for some weeks.

With respect to Dicentra it is really pretty to watch the Humble Bees sucking first on one or the other side of the several flowers; with their hind legs resting on the crests of the hood formed by the inner united petals they push it to opposite side of flower, & the straight pistil is rubbed against their abdomens & inner side of thighs, which are *white with pollen* from the several flowers. It is impossible but what the individuals of Dicentra must be largely crossed. Your Adlumia has not flowered with me yet. In Fumaria & Corydalis we have another structure, viz nectary on one side & here the pistil bends so that the 2 stigmas are presented in the gangway to the one nectary; & the hood slips off easiest in opposite direction, instead of equally easily to either side.

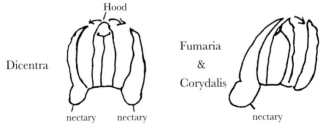

Indeed, in Corydalis lutea it almost springs off, & the pistil decidedly springs towards the nectary-bearing petal. I have observed only 6 Fumariaceæ, & I wish that I knew whether the rule was general; for I must believe that the structure of these flowers is related directly to the visits of Bees.

I *suspect* from my own few observations that the following rule may be generalised (& I sh^d. much like to know whether it is true) that when honey is secreted on one point of circle of corolla, the pistil *if it bends*, always bends so that the stigmas, when mature, lie in the gangway to nectary. Thus in Columbine where there is a circle of nectaries, the stigmas are straight; in Aquilegia grandiflora where there is one nectary, the stigmas are rectangularly bent so that every Bee (as I this day saw) brushes over them in extracting the honey.—

It is very unlikely, but if by any chance you have my little sketch of my notions of "natural Selection" & would see whether it or my letter bears any date, I sh^d. be very much obliged. Why I ask this, is as follows. M^r. Wallace who is now exploring New Guinea, has sent me an abstract of the same theory, most curiously coincident even in expressions. And he could never have heard a word of my views. He directed me to forward it to Lyell.— Lyell who is acquainted with my notions consulted with Hooker, (who read a dozen years ago a *long* sketch of mine written in 1844) urged me with much kindness not to let myself to be quite forestalled & to allow them to publish with Wallace's paper an abstract of mine; & as the only very brief thing which I had written out was a copy of my letter to you, I sent it and, I believe, it has just been read, (though never written, & not fit for such purpose) before the Linnean Soc^y.; & this is the reason, why I sh^d. be glad of the date. But do not hunt for it, as I am sure it was written in September, October or November of last year.—

I have troubled you with a long story on this head; so pray forgive me & believe me | Yours very sincerely | C. Darwin.— ...

To J. D. Hooker   13 [July 1858]

Miss Wedgwoods | Hartfield | Tunbridge Wells
13th.

My dear Hooker

Your letter to Wallace seems to me perfect, quite clear & most courteous. I do not think it could possibly be improved & I have today forwarded it with a letter of my own.—

I always thought it very possible that I might be forestalled, but I fancied that I had grand enough soul not to care; but I found myself mistaken & punished; I had, however, quite resigned myself & had written half a letter to Wallace to give up all priority to him & sh^d. certainly not have changed had it not been for Lyell's &

yours quite extraordinary kindness. I assure you I feel it, & shall not forget it.

I am **much** *more* than satisfied at what took place at Linn. Soc[y]— I had thought that your letter & mine to Asa Gray were to be only an appendix to Wallace's paper.—

We go from here in few days to sea-side, probably Isle of Wight & on my return (after a battle with Pigeon skeletons) I will set to work at abstract, though how on earth I shall make anything of an abstract in 30 pages of Journal I know not, but will try my best. ...

You cannot imagine how pleased I am that the notion of Natural Selection has acted as a purgative on your bowels of immutability. Whenever naturalists can look at species changing as certain, what a magnificent field will be open,—on all the laws of variation,—on the genealogy of all living beings,—on their lines of migration &c &c. ...

Farewell my dear kind friend— Yours affect[ly] | C. Darwin ...

[Immediately on his return to Down, CD set to work on the abstract, which eventually grew to book length; it was published in late 1859 as *On the origin of species by means of natural selection, or the preservation of favoured races in the struggle for life.*]

To T. C. Eyton   4 October [1858]

Down Bromley Kent
Oct 4th

Dear Eyton

... What a splendid collection of skeletons you have, & how many good irons you have in the fire, for I see that you are, also, going to publish a Book on the Oyster.—

I will carefully keep your letter with a list of the skeletons: at some future time the loan of some of them would be invaluable to me. I have done domestic Pigeons-skeletons, & a monograph on their history, variation &c &c.— I am so ignorant, that I do not even know the names of many bones; & I am going to take them soon to Falconer to get a little rudimentary knowledge. My notes are only 4 or 5 pages, & if I had them copied out, would you object to read them & give me your criticisms: I could, also, easily send the few bones, which show any diversity, & then you could best judge of accuracy of my few remarks. But I must first get a lesson from Faloner as soon as he returns to town.

My plans of publication are all changed, for owing to advice of Lyell & Hooker I am preparing an abstract of all my conclusions to be published as small book or read before Linn: Society, & this will for some months stop my regular work. The work is too great for me, but if I live I will finish it: indeed three-fourth is done.

With every good wish & my thanks, believe me, dear Eyton | Yours very sincerely | Ch. Darwin.

As you are a great nimrod, I wish you could tell me what colours of sire & dam will ever throw a dun coloured horse. By dun I mean a cream-colour mixed with brown or bay. I have asked scores of people, & cannot find out. Nor can I find out what is colour of a colt when born, which will turn into a Dun.—

As I am asking questions I will ask, did you ever see Ass with a double shoulder-stripe on both shoulders? Col. Ham. Smith says he has heard of such.—

To W. E. Darwin   15 [October 1858]

Down.
15th

My dear William.

You sent us a fine long letter, & we were uncommonly glad to hear that you were established. You had a precious hard day's work the first.— You are over the rooms which my cousin W. D. Fox had & in which I have spent many a pleasant hour.— I was in old court, middle stair-case, on right-hand on going into court, up one flight, right-hand door & capital rooms they were. If you find you do not like your rooms you could change another year.

I sh$^{\text{d.}}$ like to know whether my old gyp, Impey is still alive; if so please see him, & say that I enquired after him.—

I shall go up to London for a day on Tuesday & will then consult my Bankers about your affairs.— Did you pay for furniture; if not, ask whether the Cambridge tradesmen object to cheques on London Bankers: I sh$^{\text{d.}}$ much like to know this, as guide whether you had better open account with Union Bank or whether have money placed at some Cambridge Bank.— I wonder whether you could think of anyone, to ask this—

I am very glad that you like King's— it used to be a great pleasure to me.— You have to see the beautiful pictures in the Fitzwilliam. The backs of the Colleges (N.B *not* colle**d**ges as some people spell it) are indeed beautiful; I do not think there is anything in Oxford to equal them.— Remember to let me know in good time before you run short of money & do, I earnestly beg you, keep accounts *carefully*, & which, as far as I am concerned, shall be quite private.— ... Good Bye | My dear old man | Yours affect | C. Darwin ...

To J. D. Hooker   20 [October 1858]

Down Bromley Kent

20th

My dear Hooker

... I have written my paper for Gardeners Chronicle on fertilisation of Leguminous plants, & shall send it in a week or two; if Lindley does not like it I will send it to Linnean. I sh^d. very much like to hear what impression the facts make on you.—

I have been a little vexed at myself at having asked you not "to pronounce too strongly against nat. selection". I am sorry to have bothered you, though I have been much interested by your note in answer. I wrote the sentence without reflexion. But the truth is that I have so accustomed myself, partly from being quizzed by my non-naturalist relations, to expect opposition & even contempt, that I forgot for the moment that you are the one living soul from whom I have constantly received sympathy. Believe that I never forget for even a minute how much assistance I have received from you.— You are quite correct that I never even suspected that my speculations were a "jam-pot" to you: indeed I thought, until quite lately, that my M.S. had produced no effect on you & this has often staggered me. Nor did I know that you had spoken in general terms about my work to our friends, excepting to dear old Falconer, who some few years ago once told me that I should do more mischief than any ten other naturalists would do good, & that I had half-spoiled you already! All this is stupid egotistical stuff, & I write it only because you may think me ungrateful for not having valued & understood your sympathy; which God knows is not the case.—

It is an accursed evil to a man to become so absorbed in any subject as I am in mine.

I was in London yesterday for a few hours with Falconer, & he gave me a magnificent lecture on age of man. We are not upstarts; we can boast of a pedigree going far back in time coeval with extinct species. He has grand fact of some, large molar tooth in Trias.—

I am quite knocked up & am going next Monday to revive under water-cure at Moor Park.

My dear Hooker | Yours affectionately | C. Darwin ...

To Herbert Spencer   25 November [1858]

Down Bromley Kent
Nov. 25

Dear Sir

I beg permission to thank you sincerely for your very kind present of your Essays.— I have already read several of them with much interest. Your remarks on the general argument of the so-called Development Theory seem to me admirable. I am at present preparing an abstract of a larger work on the changes of species; but I treat the subject simply as a naturalist & not from a general point of view; otherwise, in my opinion, your argument could not have been improved on & might have been quoted by me with great advantage.

Your article on Music has also interested me much, for I had often thought on the subject & had come to nearly the same conclusion with you, though unable to support the notion in any detail. Furthermore by a curious coincidence Expression has been for years a favourite subject with me for *loose* speculation, & I most entirely agree with you that all expression has some biological meaning.—

I hope to profit by your criticisms on style, & with my best thanks, I beg leave to remain | Dear Sir | Yours truly obliged | C. Darwin

1859

To A. R. Wallace   25 January [1859]

Down Bromley Kent
Jan. 25th

My dear Sir

I was extremely much pleased at receiving three days ago your letter to me & that to Dr Hooker. Permit me to say how heartily I admire the spirit in which they are written. Though I had absolutely nothing whatever to do in leading Lyell & Hooker to what they thought a fair course of action, yet I naturally could not but feel anxious to hear what your impression would be. I owe indirectly much to you & them; for I almost think that Lyell would have proved right & I sh^d. never have completed my larger work, for I have found my abstract hard enough with my poor health, but now thank God I am in my last chapter, but one. My abstract will make a small vol. of 400 or 500 pages.— Whenever published, I will of course send you a copy, & then you will see what I mean about the part which I believe Selection has played with domestic productions. It is a very different part, as you suppose, from that played by "Natural Selection".— ...

I am glad to hear that you have been attending to Bird's nest; I have done so, though almost exclusively under one point of view, viz to show that instincts vary, so that selection could work on & improve them. Few other instincts, so to speak, can be preserved in a museum—

Many thanks for your offer to look after Horses stripes; if there are any Donkey's pray add them.—

I am delighted to hear that you have collected Bees' combs; when next in London I will enquire of F. Smith & Mr Saunders. This is an especial hobby of mine, & I think I can throw light on subject.— If you can collect duplicates at no very great expence, I sh^d. be glad of specimens for myself with some Bees of each kind.— Young growing & irregular combs, & those which have not had pupæ are most valuable for measurements & examination: their edges sh^{d}. be well protected against abrasion.—

Everyone whom I have seen has thought your paper very well written & interesting. It puts my extracts, (written in 1839 now just 20 years ago!) which I must say in apology were never for an instant intended for publication, in the shade.

You ask about Lyell's frame of mind. I think he is somewhat staggered, but does not give in, & speaks with horror often to me, of what a thing it would be & what a job it would be for the next Edition of Principles, if he were "*per*verted".— But he is most candid & honest & I think will end by being *per*verted.— D^{r}. Hooker has become almost as heteredox as you or I.—and I look at Hooker as **by far** the most capable judge in Europe.—

Most cordially do I wish you health & entire success in all your pursuits & God knows if admirable zeal & energy deserve success, most amply do you deserve it.

I look at my own career as nearly run out: if I can publish my abstract & perhaps my greater work on same subject, I shall look at my course as done.

Believe me, my dear Sir | Yours very sincerely | C. Darwin

To W. D. Fox   [12 February 1859]

Moor Park, Farnham | Surrey
Saturday

My dear Fox

It is long since we have had any communication, so I am tempted to send you a scrap to extract a scrap from you, with news about yourself.— I have been extra bad of late, with the old severe vomiting rather often & much distressing swimming of the head; I have been here a week & shall stay another & it has already done me good. I am taking Pepsine, ie the chief element of the gastric juice, & I think it does me good & at first was charmed with it. My abstract is the cause, I believe of the main part of the ills to which my flesh is heir to; but I have only two more chapters & to correct all, & then I shall be a comparatively free man.— I have had the great satisfaction of converting Hooker & I believe Huxley & I think Lyell is much staggered. . . .

William is very happy at Cambridge & he has changed into my old rooms & has taken my old engravings & with old Impey, it must be a sort of resurrection.

Let me some time have a note telling me about yourself & be-
longings

My dear Fox | Ever yours most truly | C. Darwin

To J. D. Hooker   2 March [1859]

Down. Bromley Kent
March 2^d

My dear Hooker

... I have finished my abstract of Ch. on Geograph. Distrib. as
bearing on my subject.— I should like you much to read it; but I
say this, believing that you will not do so, if, as I believe to be the
case, you are extra busy. On my honour I shall not be mortified &
I **earnestly beg** you not to do it, if it will bother you.— I want it,
because I here feel especially unsafe & errors may have crept in. Also
I sh^d. much like to know what parts you will *most vehemently* object to; I
know we do, & must, differ widely on several heads. Lastly I sh^d. like
particularly to know, whether I have taken anything from you, which
you w^d like to retain for first publication; but I think I have chiefly
taken from your published works; & though I have several times in
this Ch. & elsewhere acknowledged your assistance, I am aware that
it is not possible for me in this abstract to do it sufficiently.— ...

Yours affect | C. Darwin ...

To Charles Lyell   28 March [1859]

Down Bromley Kent
March 28th

My dear Lyell

If I keep decently well I hope to be able to go to press with my
volume early in May. This being so, I want much to beg a little
advice from you.— From an expression in Lady Lyell's note, I fancy
that you have spoken to Murray. Is it so? and is he willing to publish
my Abstract? If you will tell me whether anything & what has passed,
I will then write to him: does he know at all subject of Book?

Secondly can you advise me, whether I had better state what terms
of publication I sh^d. prefer or first ask him to propose terms. And
what, do you think, would be fair terms for an Edition? share profits
or what?

Lastly, will you be so very kind as to look at enclosed title & give
me your opinion & any criticisms: you must remember that if I have
health & it appears worth doing, I have a much larger & full book

on same subject nearly ready. My abstract will be about 500 pages of size of your first Edition of Elements of Geology.—

Pray forgive me troubling you with above queries; & you shall have no more trouble on subject.

I hope the world goes well with you, & that you are getting on with your various works.

I am working very hard for me, & long to finish & be free & try to recover some health.

My dear Lyell | Ever yours | C. Darwin . . .

P.S. | Would you advise me to tell Murray that my Book is not more *un*orthodox, than the subject makes inevitable. That I do not discuss origin of man.— That I do not bring in any discussions about Genesis &c, & only give facts, & such conclusions from them, as seem to me fair.—

Or had I better say *nothing* to Murray, & assume that he cannot object to this much unorthodoxy, which in fact is not more than any Geological Treatise, which runs slap counter to Genesis.

[Enclosure]

An abstract of an Essay
on the
Origin
of
Species and Varieties
Through Natural Selection
by
Charles Darwin M. A
Fellow of the Royal, Geological & Linn. Soc.^{y}.

———

London
&c &c &c &c
1859

———

To Charles Lyell   30 March [1859]

Down Bromley Kent
March 30th

My dear Lyell

You have been uncommonly kind in all you have done.— You not only have saved me much trouble & some anxiety, but have done all, *incomparably* better than I could have done it— I am much pleased at all you say about Murray.— I will write either today or tomorrow to him & will send shortly a large bundle of M.S. but unfortunately I cannot for a week, as the three first chapters are in three copyists' hands—

I am sorry about Murray objecting to term abstract as I look at it as only possible apology for *not* giving References & facts in full.—but I will defer to him & you.—

I am, also, sorry about term "Natural Selection", but I hope to retain it with Explanation, somewhat as thus,—"Through Natural Selection or the preservation of favoured races"

Why I like term is that it is constantly used in all works on Breeding, & I am surprised that it is not familiar to Murray; but I have so long studied such works, that I have ceased to be a competent judge.

I again most truly & cordially thank you for your really valuable assistance.—

Yours most truly | C. Darwin . . .

To J. D. Hooker   2 April [1859]

Down Bromley Kent
April 2^{d}

My dear Hooker

Very many thanks for your letter of caution about Murray. I wrote to him & gave him the Headings of chapters, and told him he could not have M.S for 10 days or so., and this morning I receive a letter, offering me handsome terms & agreeing to publish without seeing M.S! So he is eager enough; I think, I sh^{d} have been cautious anyhow, but owing to your letter, I have told him most *explicitly*, that I accept his offer solely on condition, that after he has seen part or all M.S. he has **full** power of retracting.— You will think me presumptuous, but I think my book will be popular to a certain extent, enough to ensure heavy loss amongst scientific & semi-scientific men: why I think so is because I have found in conversation so great & surprising interest amongst such men & some o-scientific men on subject; & all my chapters are not *nearly* so dry & dull as that which you have read on Geographical Distribution.— Anyhow Murray ought to be the best judge, & if he chooses to publish it, I think I may wash my hands of all responsibility.— And he made very good bargain for my Journal I am sure my friends, ie Lyell

& you have been *extraordinarily* kind in troubling yourselves on the matter. . . .

    I am tired, so no more | My dear Hooker
Yours affect^y | C. Darwin . . .

To John Murray   2 April [1859]

                              Down Bromley Kent
                                          April 2^d
My dear Sir

    I am much obliged for your note, & accept with pleasure your offer. But I feel bound for your sake (& my own) to say in clearest terms, that if after looking over part of M.S. you do not think it likely to have a remunerative sale, I **completely** & explicitly free you from your offer. But you will see that it would be a stigma on my work for you to advertise it, & then not publish it. My volume cannot be mere light reading, & some parts must be dry & some rather abstruse; yet *as far I can judge perhaps very falsely*, it will be interesting to all (& they are many) who care for the curious problem of the origin of all animate forms. . . .

    With my thanks & hearty wishes that you may not be disappointed in work, if published by you, pray believe me, My dear Sir | Yours very sincerely | C. Darwin . . .

    P.S. I would add that it is impossible for you or anyone to judge of real merit of my Book, without reading the whole, as the whole is one long argument

To A. R. Wallace   6 April 1859

                              Down Bromley Kent
                                     April 6 /59
My dear Mr Wallace

    I this morning received your pleasant & friendly note of Nov. 30th. The first part of my M.S is in Murray's hands to see if he likes to publish it. There is no preface, but a short Introduction, which must be read by everyone, who reads my Book. The second Paragraph in the Introduction, I have had copied *verbatim* from my foul copy, & you will, I hope, think that I have fairly noticed your paper in Linn. Transacts— You must remember that I am now publishing only an Abstract & I give no references.— I shall of course allude to your paper on Distribution; & I have added that I know from correspondence that your explanation of your law is the same as that which I offer.— You are right, that I came to conclusion that Selection was the principle of change from study of domesticated productions; & then reading Malthus I saw at once how to apply this principle.— Geographical Distrib. & Geological relations of extinct to recent inhabitants of S. America first led me to subject. Especially case of Galapagos Isl^{ds}.—

I hope to go to press in early part of next month.— It will be small volume of about 500 pages or so.— I will of course send you a copy. I forget whether I told you that Hooker, who is our best British Botanist & perhaps best in World, is a *full* convert, & is now going immediately to publish his confession of Faith; & I expect daily to see the proof-sheets.— Huxley is changed & believes in mutation of species: whether a *convert* to us, I do not quite know.— We shall live to see all the *younger* men converts. My neighbour & excellent naturalist J. Lubbock is enthusiastic convert.

I see by Nat. Hist notices that you are doing great work in the Archipelago; & most heartily do I sympathise with you. For God sake take care of your health. There have been few such noble labourers in the cause of Natural Science as you are.

Farewell, with every good wish | Yours sincerely | C. Darwin

P.S. | You cannot tell how I admire your spirit, in the manner in which you have taken all that was done about publishing our papers. I had actually written a letter to you, stating that I would *not* publish anything before you had published. I had not sent that letter to the Post, when I received one from Lyell & Hooker, *urging* me to send some M.S. to them, & allow them to act as they thought fair & honourably to both of us. & I did so.—

To J. D. Hooker   12 [April 1859]

Down Bromley Kent
12th

My dear Hooker

I **have** the old M.S, otherwise the loss would have killed me!¹ The worst is now that it will cause delay in getting to press, & *far worst* of all I lose all advantage of you having looked over my chapter, except the third part returned.— I am very sorry Mrs Hooker took the trouble of copying the two pages. ...

I would advise you to be cautious about stating so broadly (I thought that you perhaps knew of distinct cases unknown to me) about species not varying for **many** generations & then **suddenly** varying. To a certain extent I *quite* believe it; ie that a plant will not vary until after some few generations (perhaps dozen or so) & then will begin to vary *possibly* suddenly, more likely gradually. But even my belief in this is grounded on very few facts.— I believe another & very distinct explanation may be given of a sort of current belief in the doctrine, viz that variations are often not attended to, & till they are attended to & accumulated, they make no show.—

Ever yours | C. Darwin

I have worked this notion up in (as it seems to me) an important manner in my Ch. on Domestication of Animals & Plants.

To A. R. Wallace   9 August 1859

Down Bromley Kent
Augst. 9th. 1859

My dear Mr Wallace

I received your letter & memoir on the 7th & will forward it tomorrow to Linn. Soc^y. But you will be aware that there is no meeting till beginning of November. Your paper seems to me *admirable* in matter, style & reasoning; & I thank you for allowing me to read it.[2] Had I read it some months ago I sh^d. have profited by it for my forthcoming volume.— But my two chapters on this subject are in type; & though not yet corrected, I am so wearied out & weak in health, that I am fully resolved not to add one word & merely improve style. So you will see that my views are nearly the same with yours, & you may rely on it that not one word shall be altered owing to my having read your ideas. . . .

I differ **wholly** from you on colonisation of *oceanic* islands, but you will have **everyone** else on your side. I quite agree with respect to all islands not situated far in ocean. I quite agree on little occasional intermigration between lands when once pretty well stocked with in-habitants, but think this does not apply to rising & ill-stocked islands. Are you aware that *annually* birds are blown to Madeira, to Azores, (& to Bermuda from America).— I wish I had given fuller abstract of my reasons for not believing in Forbes' great continental extensions; but it is too late, for I will alter nothing. I am worn out & must have rest.—

Owen, I do not doubt, will bitterly oppose us; but I regard this very little; as he is a poor reasoner & deeply considers the good opinion of the world, especially the aristocratic world.—

Hooker is publishing a grand Introduction to Flora of Australia & goes the whole length.— I have seen proofs of about half.—

With every good wish. Believe me | Yours very sincerely | C. Darwin | Excuse this brief note, but I am far from well.—

To Charles Lyell   20 September [1859]

Down Bromley Kent
Sept 20th

My dear Lyell

You once gave me intense pleasure, or rather delight, by the way you were interested, in a manner I never expected, in my Coral-reef notions; & now you have again given me similar pleasure by the manner you have noticed my Species work.[3] Nothing could be more satisfactory to me, & I thank you for myself, & even more for the subject-sake, as I know well that sentence will make many fairly consider the subject, instead of ridiculing it. Although your previously felt doubts on the immutability of species, may have more influence in converting you (if you be converted) than my Book; yet as I regard your verdict as far more important in my own eyes & I believe in eyes of world than of any other dozen men, I am naturally very anxious about it. Therefore, let me beg you to keep your mind open till you receive (in perhaps a fortnights time) my latter chapters which are the most important of all on the favourable side. The last chapter which sums up & balances in a mass all the arguments contra & pro, will, I think, be useful to you.—[4]

I cannot too strongly express my conviction of the general truth of my doctrines, & God knows I have never shirked a difficulty.— I am foolishly anxious for your verdict. Not that I shall be disappointed if you are not converted; for I remember the long years it took me to come round; but I shall be most deeply delighted if you do come round, especially if I have a fair share in the conversion. I shall then feel that my career is run, & care little whether I ever am good for anything again in this life. ...

With cordial thanks for your splendid notice of my Book | Believe me, my dear Lyell | Your affectionate disciple | Charles Darwin ...

To Leonard Jenyns   13 November [1859]

Wells Terrace | Ilkley, Otley | Yorkshire
Nov. 13th

My dear Jenyns

I must thank you for your very kind note forwarded to me from Down.— I have been much out of health this summer & have been

hydropathising here, for last six weeks with very little good as yet.—
I shall stay here for another fortnight at least.

Please remember that my Book is only an abstract & very much
condensed & to be at all intelligible must be carefully read. I shall
be very grateful for any criticisms. But I know perfectly well that
you will not at all agree with the lengths which I go. It took long
years to convert me.— I may of course be egregiously wrong; but
I cannot persuade myself that a theory which explains (as I think
it certainly does) several large classes of facts, can be wholly wrong;
notwithstanding the several difficulties which have to be surmounted
somehow, & which stagger me even to this day.

I wish that my health had allowed me to publish in extenso; if I
ever get strong enough I will do so, as the greater part is written out,
& of which M.S. the present volume is an abstract.—

I fear this note will be almost illegible; but I am poorly & can
hardly sit up.

Farewell with thanks for your kind note & pleasant remembrances
of good old days | Yours very sincerely | C. Darwin

To T. H. Huxley   24 [November 1859]

Ilkley Wells House | Otley, Yorkshire
24th

My dear Huxley

I have heard from Murray today that he sold whole Edition of my
Book on first day, & he wants another instantly, which confounds me,
as I can make hardly any corrections.[5] But a friend writes to me that
it ought to be Geoffroy DE St. Hilaire: my memory says *no*. Will you
turn to a title-page & tell me soon & forgive me asking this trouble.

Remember how deeply I wish to know your general impression
of the truth of the theory of Natural Selection.—only a short note—
at some future time if you have any lengthy criticisms, I sh^d be
infinitely grateful for them. You must know well how highly I value
your opinion.—

In Haste, for I am bothered to death by this new Edition | Ever
yours | C. Darwin

To Charles Lyell   [10 December 1859]

Down Bromley Kent
Saturday

My dear Lyell

... I have [had a] very long interview with Owen, which perhaps you would like to hear about, **but please repeat nothing**. Under garb of great civility, he was inclined to be most bitter & sneering against me. Yet I infer from several expressions, that *at bottom* he goes immense way with us.— He was quite savage & crimson at my having put his name with defenders of immutability. When I said that was my impression & that of others, for several had remarked to me, that he would be dead against me: he then spoke of his own position in science & that of all the naturalists in London, "with your Huxleys", with a degree of arrogance I never saw approached. He said to effect that my explanation was best ever published of manner of formation of species. I said I was very glad to hear it. He took me up short, "you must not at all suppose that I agree with in all respects".— I said I thought it no more likely that I sh^d be right on nearly all points, than that I sh^d toss up a penny & get heads twenty times running.

I asked him which he thought the weakest parts,— he said he had no particular objection to any part.— He added in most sneering tone if I must criticise I sh^d say "we do not want to know what Darwin believes & is convinced of, but what he can prove".— I agreed most fully & truly that I have probably greatly sinned in this line, & defended my general line of argument of inventing a theory, & seeing how many classes of facts the theory would explain.— I added that I would endeavour to modify the "believes" & "convinceds". He took me up short,—"You will then spoil your book, the **charm of**(!) it is that it is Darwin himself".— He added another objection that the book was too "teres atque rotundus",[6] —that it explained everything & that it was improbable in highest degree that I sh^d succeed in this". I quite agree with this rather queer objection, & it comes to this that my book must be very bad or very good.— Lastly I thanked him him for Bear & Whale criticism, & said I had struck it out.—[7] "Oh have you, well I was more struck with this than any other passage; you little know of the remarkable & essential relationship between bears & whales".—

I am to send him the reference, & by Jove I believe he thinks a sort of Bear was the grandpapa of Whales! I do not know whether I have wearied you with these details which do not repeat to any one.— We parted with high terms of consideration; which on reflexion I am almost sorry for.— He is the most astounding creature I ever encountered.

Farewell my dear Lyell | Yours most gratefully | C. Darwin

I have heard by round about channel that Herschel says my Book "is the law of higgledy-pigglety".— What this exactly means I do not know, but it is evidently very contemptuous.— If true this is great blow & discouragement.

To T. H. Huxley 25 December [1859]

Down Bromley Kent
Dec. 25th

My dear Huxley

One part of your note has pleased me so much that I must thank you for it. Not only Sir H[enry]. H[olland]. but several others have attacked me about analogy leading to belief in one primordial **created** form. (By which I mean only that we know nothing as yet how life originates). I thought I was universally condemned on this head.— But I answered, that though perhaps it would have been more prudent not to have put it in, I would not strike it out, as it seemed to me probable & I give it on no other grounds.— You will see in your mind the kind of arguments which made me think it probable; & no one fact had so great an effect on me, as your most curious remarks on the apparent homologies of the heads of Vertebrata & Articulata.—

You have done a real good turn in the Agency business (I never before heard of a hard-working unpaid agent besides yourself) in talking with Sir H. H; for he will have great influence over many. He floored me from my ignorance about bones of Ear, &I made a mental note to ask you what the facts were.—

With hearty thanks & real admiration for your generous zeal for the subject.— | Yours most truly | C. Darwin . . .

To T. H. Huxley 28 December [1859]

Down Bromley Kent
Dec. 28th

My dear Huxley

Yesterday Evening when I read the Times of previous day I was amazed to find a splendid Essay & Review of me. Who can the author be? I am intensely curious. It included a eulogium of me, which quite touched me, though I am not vain enough to think it all deserved.— The Author is a literary man & German scholar.— He has read my Book very attentively; but what is very remarkable, it seems that he is a profound naturalist. He knows my Barnacle book, & appreciates it too highly.— Lastly he writes & thinks with quite uncommon force & clearness; & what is even still rarer his writing is seasoned with most pleasant wit. We all laughed heartily over some of the sentences. I was charmed with those unreasonable mortals who know everything all thinking fit to range themselves on our side Who can it be? Certainly I should have said that there was only one man in England who could have written this Essay & that *you* were the man. But I suppose I am wrong, & that there is some hidden genius of great calibre. For how could you influence Jupiter Olympus[8] & make him give $3\frac{1}{2}$ columns to pure science? The old Fogies will think the world will come to an end.—

Well whoever the man is, he has done great service to the cause, far more than by a dozen Reviews in common periodicals. The grand way he soars above common religious prejudices, & the admission of such views into the Times, I look at as of the highest importance, quite independently of the mere question of species. If you should happen to be *acquainted* with the author for Heaven-sake tell me who he is.—

My dear Huxley | Yours most sincerely | C. Darwin . . .

Notes

Introduction

[1] See Bibliographical note (p. 237) for this and other Darwin works mentioned in the letters.

[2] Ralph Colp Jr, in *To be an invalid* (University of Chicago Press, 1977) advanced a psychological view of CD's symptoms and discussed other theories of their cause. Since then, John Bowlby has proposed a new hypothesis, based on the effects of childhood bereavement, in *Charles Darwin. A new life* (Norton, 1990), and Fabienne Smith has offered the hypothesis that CD 'suffered from multiple allergy arising from a dysfunctional immune system.' (*Journal of the History of Biology* 23 (1990): 443–59; 25 (1992): 285–306).

[3] *Correspondence* 2: 431.

[4] *Correspondence* 2: 107.

[5] A. R. Wallace. On the law which has regulated the introduction of new species. *Annals and Magazine of Natural History* 2d ser. 16 (1855): 184–96.

[6] Charles Darwin and A. R. Wallace. On the tendency of species to form varieties, and on the perpetuation of varieties and species by means of natural selection. [Read 1 July 1858.] *Journal of the Proceedings of the Linnean Society (Zoology)* 3 (1859): 45–62.

Edinburgh

[1] *Der Freischütz*, an opera by Carl Maria von Weber.

[2] A novel by Thomas Henry Lister. *Granby*. 3 vols. London, 1826.

[3] A popular Tory weekly, notorious for scandal-mongering. CD and Erasmus read it for the scandal and sent it home as a tease. The Darwins were Whigs.

Cambridge

[1] J. F. Stephens. *Illustrations of British entomology*. London, 1827–46.

[2] In a letter to Fox (24 December 1828), CD referred to Fanny as the 'prettiest, plumpest, Charming personage that Shropshire posseses'.

3 The examination taken by undergraduates in their second year.
4 CD may have intended to write 'Severn formations'; Shrewsbury is located on the River Severn.
5 Alexander von Humboldt. *Personal narrative of travels to the equinoctial regions of the New Continent during the years 1799–1804.* London, 1814–29.

The Offer

1 F. Darwin. FitzRoy and Darwin, 1831–36. *Nature* 88 (1912): 547–8.

The Voyage: South America – East Coast

1 CD had heard from his sisters of Fanny Owen's sudden engagement to Robert Myddelton Biddulph.
2 Alcide Charles Victor Dessalines d'Orbigny travelled throughout South America between 1826 and 1833, collecting specimens for the Muséum d'Histoire Naturelle.
3 *Paradise lost* 4.799–800. CD had a copy of Milton's poems with him on the voyage.
4 Patrick Syme. *Werner's nomenclature of colours.* Edinburgh, 1814. The work was a handbook of standard colours for describing natural history specimens.

The Voyage: South America – West Coast

1 J. V. F. Lamouroux. *Exposition méthodique des genres de l'ordre des Polypiers.* Paris, 1821.
2 John Narbrough. *An account of several late voyages. . .towards the Streights of Magellan* etc. London, 1694.
3 Claude Gay. Aperçu sur les recherches d'histoire naturelles faites dans l'Amérique du Sud, et principalement dans le Chili, pendant les années 1830 et 1831. *Annales des Sciences Naturelles* 28 (1833): 369–93.
4 Viscount Melbourne succeeded Lord Grey as prime minister.
5 The London, Birmingham, Shrewsbury mail coach.
6 Dated from the arrival of the *Beagle* at Callao, the port for Lima (19 July 1835). July is a mistake for August.

Homeward Bound

1 No CD letter from the Galápagos has been found.
2 Augustus Earle. *A narrative of nine months residence in New Zealand, in 1827. . .* London, 1832. [Reprinted: Cambridge University Press, 1966.]

[3] Six months later, on finding strong feelings at Cape Town against missionaries, FitzRoy and CD published a defence of their work in the *South African Christian Recorder* (see *Collected papers* 1: 19–38).

[4] This was the generally held view. CD's theory, published in *Coral reefs* (1842), held that the reefs were formed by the upward growth of coral during the gradual subsidence of the sea-bed adjacent to volcanic islands.

[5] Sedgwick wrote to Samuel Butler, headmaster of Shrewsbury School: 'He ... has already sent home a Collection above all praise.— It was the best thing in the world for him that he went out on the Voyage of Discovery—There was some risk of his turning out an idle man: but his character will now be fixed, & if God spare his life, he will have a great name among the Naturalists of Europe' (in a letter from Susan Darwin, 22 November 1835, *Correspondence* 1: 469).

1837

[1] In his presidential address to the Geological Society, Lyell, besides mentioning CD's 'Observations of proofs of recent elevation on the coast of Chili' (*Collected papers* 1: 41–3), reported briefly on the 'most striking results' of Richard Owen's examination of CD's South American fossil collection. 'The Missionary paper' was a defence of the work of missionaries, jointly written by FitzRoy and CD for publication in the *South African Christian Recorder* 2 (1836): 221–38; *Collected papers* 1: 19–38.

[2] The only version of this letter known to exist is a copy held in the Darwin Archive, Cambridge University Library.

[3] The passage is in Erasmus Darwin's *Zoonomia* (London, 1794–6) 1: 158: '[rooks] evidently distinguish, that the danger is greater when a man is armed with a gun.' CD had become interested in the inheritance of 'acquired instincts' because of the tameness of the animals on the Falkland and Galápagos islands (see *Voyage of the Beagle*, p. 290).

[4] William Ellis. *Polynesian researches, during a residence of nearly six years in the South Sea islands.* 2 vols. London, 1829.

[5] Most of the requested information is cited in chapter 21 of *Voyage of the Beagle*, in the discussion of the 'mysterious' causes of extinction of native populations. 'Wherever the European has trod, death seems to pursue the aboriginal' (pp. 321–2).

⁶ Andrew Smith had informed CD that South Africa supported many large mammals in spite of its relatively dry climate (*Voyage of the Beagle*, pp. 99–101).

⁷ William Whewell, president of the Geological Society, had asked CD to accept appointment to a secretaryship of the society.

1838

¹ A 'goose' was a Darwin family term for an intimate conversation.

1839–1843

¹ The Darwins' first child, William Erasmus Darwin, was born on 27 December 1839.

² A mean old usurer in Charles Dickens's *Nicholas Nickleby* (London, 1839).

³ Anne Elizabeth Darwin was born on 2 March 1841.

⁴ A character in two plays: John Vanbrugh's *The relapse* (1696) and Richard Brinsley Sheridan's *A trip to Scarborough* (1777) (*Oxford companion to English literature*).

⁵ Hooker had been assistant surgeon on HMS *Erebus* and botanist to the Antarctic expedition under James Clark Ross (1839–43). He returned to England in September 1843.

1844

¹ CD first formulated his theory of natural selection in the autumn of 1838 (*Notebooks* D: 134e–5e).

² Charles Lyell. *Principles of geology: or, the modern changes of the earth and its inhabitants, considered as illustrative of geology.* 6th ed. 3 vols. London, 1840.

³ Leonard Jenyns, ed. *The natural history of Selbourne by the late Rev. Gilbert White, M.A. A new edition with notes.* London, 1843.

1845–1846

¹ Henrietta Emma Darwin was born on 25 September 1843.

² George Howard Darwin was born on 9 July 1845.

³ CD was preparing a second edition of the *Voyage of the Beagle*.

⁴ The author of the anonymous *Vestiges of the natural history of creation* (London, 1844) was Robert Chambers.

⁵ William Herbert. Local habitation and wants of plants. *Journal of the Horticultural Society of London* 1 (1846): 44–9.

6 CD's *South America* (1846).

7 J. D. Hooker. *Flora Antarctica*. Pt 1 of *The botany of the Antarctic voyage of H.M. Discovery Ships Erebus and Terror in the years 1839–1843, under the command of Captain Sir James Clark Ross*. 2 vols. London, 1844–7.

1847

1 E. W. Binney. Description of the Dickinfield Sigillaria. *Journal of the Geological Society of London* 2 (1846): 390–3.

2 Elizabeth Darwin was born on 8 July 1847.

1848

1 James Smith. On recent depressions in the land. *Quarterly Journal of the Geological Society of London* 3 (1847): 234–40.

2 J. S. Henslow. *Address delivered in the Ipswich Museum, on 9th of March, 1848*. Ipswich, 1848.

3 CD discovered separate sexes in *Ibla cumingii*, in contrast to the majority of cirripedes, which are hermaphrodites (see *Living Cirripedia* (1851): 189–203).

4 Francis Darwin was born on 16 August 1848.

5 The 'hoi polloi' or 'Poll' was the undergraduate term for those who read for a 'pass degree'. CD in fact came tenth out of 178 who passed.

6 This letter was written at Erasmus Alvey Darwin's house when CD was on his way to Shrewsbury to attend his father's funeral. On his arrival in Shrewsbury, the funeral service had begun, but CD was too ill too attend.

1849

1 Charles Lyell. *A second visit to the United States of North America*. 2 vols. London, 1849.

2 The 1849 meeting of the British Association for the Advancement of Science was to be held in Birmingham.

3 The complemental males of *Scalpellum* (see *Living Cirripedia* (1851): 231–43 and 281–93).

4 J. D. Dana. *Geology*. Vol. 10 of *United States Exploring Expedition during the years 1838–1842, under the command of Charles Wilkes, U.S.N.* New York, 1849.

1850

1 Leonard Darwin was born on 15 January 1850.

1852–1854

[1] In 1845, CD had purchased some Lincolnshire farmland. John Higgins was his agent.

[2] Huxley was cataloguing the British Museum collection of Ascidia (sea-squirts).

[3] This ellipsis is Darwin's own.

[4] A character in Charles Dickens's *Bleak House* (London, 1853).

[5] The Hookers' second child, Harriet Anne, was born on 23 June 1854.

[6] M. J. Schleiden. *The plant; a biography. In a series of popular lectures.* Translated by Arthur Henfrey. London, 1848.

[7] Huxley's review of the tenth edition of Robert Chambers's anonymous *Vestiges of the natural history of creation.* In 1887, Huxley referred to this piece as 'the only review I ever have qualms of conscience about, on the ground of needless savagery' (L. Huxley ed. 1900, 1: 168). The 'great Professor' is Richard Owen.

[8] This ellipsis is Darwin's own.

1855

[1] On the front cover of a heavily annotated copy of John Cattell's 1855 catalogue of floricultural seeds, CD wrote in ink: 'Hooker suggests for sea-water | 1 Plants with wide ranges | 2 Water Plants. *Ask Cattell how I am to get any? [added] | 3 Plants with farinaceous album | 4 With fleshy d[itt]o | 5 With oily d[itt]o.' (Darwin Archive, Cambridge University Library.)

[2] CD refers to his plan to identify the different species of grass growing in the neighbourhood of Down.

[3] The text is as printed in the *Gardeners' Chronicle*; the word 'evergreens' was mistakenly printed instead of CD's intended word, 'Euonymus'.

1856

[1] Huxley's 'Lectures on general natural history', published in the *Medical Times*, contained repeated attacks on Richard Owen.

[2] 'This is the question, which has long agitated naturalists, namely whether the same species has been created once & therefore at a single point, or more than once at different points.' (*Natural selection*, p. 534.)

³ Johann Matthäus Bechstein's book on the natural history of Germany (*Gemeinnützige Naturgeschichte Deutschlands nach allen drey Reichen.* 4 vols. Leipzig, [1789–95]) was frequently cited by CD in *Natural selection.*

⁴ 'I have finished the reading of your mss. . . . I never felt so shaky about species before.' (letter from J. D. Hooker, 9 November 1856, *Correspondence* 6: 259).

1857

¹ Charles Waring Darwin was born on 6 December 1856.

² The first Austrian scientific expedition to circumnavigate the globe was announced in the *Athenæum*, 10 January 1857, p. 53. The leader of the expedition, Karl von Scherzer, requested advice to render the voyage 'efficient and fruitful in valuable results.'

³ A reference to Horace, *Satires*, 1.3.107, in which women were deemed 'the most foul cause of war'.

⁴ A. R. Wallace. On the law which has regulated the introduction of new species. *Annals and Magazine of Natural History* 2d ser. 16 (1855): 184–96. The law: 'Every species has come into existence coincident both in time and space with a pre-existing closely allied species.'

1858

¹ CD had sent Hooker his manuscript on variation in large and small genera. It is printed, with Hooker's comments, in *Natural selection*, pp. 138–67.

² Wallace had sent CD the manuscript of a paper entitled 'On the tendency of varieties to depart indefinitely from the original type'.

³ Charles Waring Darwin eventually died of scarlet fever on 28 June 1858.

⁴ CD's daughter, Henrietta Emma Darwin, aged 14.

⁵ Hooker and Charles Lyell had evidently suggested in the letters mentioned by CD that they submit Alfred Russel Wallace's paper on species together with extracts from CD's writings to the Linnean Society as a joint paper. This paper would make public CD's work of twenty years and at the same time establish proof of his independent formulation of the concept of natural selection.

1859

[1] Hooker had accidentally put the fair copy of CD's chapters on geographical distribution into the chest in which his wife kept paper for the children to draw on. In a letter to Thomas Henry Huxley, Hooker related how the children 'of course had a drawing fit ever since' so that nearly a quarter of the manuscript had vanished before Hooker came to read it (L. Huxley ed. *Life and letters of Sir Joseph Dalton Hooker* (London, 1918) 1: 495–6).

[2] Wallace's memoir was on the zoological geography of the Malay Archipelago. CD communicated the paper to the Linnean Society; it was read at the meeting of 3 November 1859.

[3] In his opening address to the geology section of the British Association for the Advancement of Science meeting in Aberdeen (September 1859), Lyell stated that CD appeared 'to have succeeded, by his investigations and reasonings, in throwing a flood of light on many classes of phenomena connected with the affinities, geographical distribution, and geological succession of organic beings, for which no other hypothesis has been able, or has even attempted, to account.' (*Athenæum*, 24 September 1859, p. 404).

[4] CD finished reading proofs of *Origin* on 1 October 1859, and received his personal copy of the volume on 2 November. He ordered presentation copies, which were sent out during the second week of November in advance of the publication date, set for 24 November.

[5] Orders for copies of *Origin* from book-dealers at John Murray's trade sale on 22 November exceeded by more than 250 the number available (1192) from the printing of 1250 copies (*Athenæum*, 26 November 1859, p. 706).

[6] Horace in the *Satires*, 2.7.86–7, described the Stoic wise man as 'totus teres atque rotundus' ('complete, polished, and round').

[7] 'In North America the black bear was seen by [Samuel] Hearne swimming for hours with widely open mouth, thus catching, like a whale, insects in the water. Even in so extreme a case as this, …, I can see no difficulty in a race of bears being rendered, by natural selection, more and more aquatic in their structure and habits, with larger and larger mouths, till a creature was produced as monstrous as a whale.' (*Origin*, p. 184).

[8] Anthony Trollope referred to *The Times* as the 'Daily Jupiter', in *The Warden* (1855) and other novels.

Biographical Register

This list includes all correspondents and most of the persons mentioned in the letters. Following the register is a list of the main biographical sources used in its compilation.

Agassiz, Jean Louis Rodolphe (Louis) (1807–73). Swiss geologist and zoologist. Professor of natural history, Neuchâtel, 1832–46. Professor of natural history, Harvard, 1847–73.

Allen, Elizabeth (Aunt Bessy) (1764–1846). Married Josiah Wedgwood II in 1792.

Babbage, Charles (1791–1871). Mathematician and pioneer in the design of mechanical computers.

Baily, John. Poulterer and dealer in live birds.

Baily, William Hellier (1819–88). Assistant geologist, Geological Survey, 1845; geologist, 1853. Palaeontologist to the Irish branch of the Geological Survey, 1857–88.

Beaufort, Francis (1774–1857). Naval officer. Hydrographer to the Admiralty, 1832–55.

Bechstein, Johann Matthäus (1757–1822). German forestry scientist and ornithologist.

Bell, Thomas (1792–1880). Dental surgeon, Guy's Hospital, London, 1817–61. Professor of zoology, King's College, London, 1836. Described the reptiles from the *Beagle* voyage.

Bentham, George (1800–84). Botanist. Pursued his botanical studies at the Royal Botanic Gardens, Kew.

Berkeley, Miles Joseph (1803–89). Clergyman and botanist. Perpetual curate of Apethorpe and Wood Newton, Northamptonshire, 1833–68.

Biddulph, Robert Myddelton (1805–72). MP for Denbighshire, 1832–5 and 1852–68. Married Fanny Owen in 1832.

Binney, Edward William (1812–81). Palaeobotanist. Solicitor in Manchester.

Birch, Samuel (1813–85). Egyptologist and archaeologist. Assistant keeper, department of antiquities, British Museum; keeper, oriental, British, and medieval antiquities, 1861–85.

Blyth, Edward (1810–73). Curator of the museum of the Asiatic Society of Bengal, 1841–62.

Bosquet, Joseph Augustin Hubert de (1814–80). Stratigrapher and palaeontologist. Pharmacist in Maastricht.

Brodie (d. 1873). The Darwin children's nurse at 12 Upper Gower Street and Down House, 1842–51.

Brongniart, Alexandre (1770–1847). French geologist. Director of the Sèvres porcelain factory, 1800–47. Professor of mineralogy, Muséum d'Histoire Naturelle, 1822.

Brooke, James (1803–68). Raja of Sarawak, Borneo, 1841–63. Appointed British commissioner and consul-general of Borneo in 1847.

Brown, Robert (1773–1858). Botanist. Librarian to Joseph Banks, 1810–20. Keeper of the botanical collections, British Museum, 1827–58.

Butler, Samuel (1774–1839). Educator and clergyman. Headmaster of Shrewsbury School, 1798–1836. Bishop of Lichfield and Coventry, 1836–9.

Bynoe, Benjamin (1804–65). Naval surgeon, 1825–63. Assistant surgeon in the *Beagle*, 1832–7; surgeon, 1837–43.

Caldcleugh, Alexander (d. 1858). Business man and plant collector in South America.

Campbell, Andrew. Superintendent of the Darjeeling station, 1840. Travelled with Joseph Dalton Hooker in Sikkim; imprisoned with Hooker by the Sikkim raja.

Candolle, Alphonse de (1806–93). Swiss botanist. Professor of botany and director of the botanic gardens, Geneva, 1835–50. Son of Augustin Pyramus de Candolle.

Candolle, Augustin Pyramus de (1778–1841). Swiss botanist. Professor of natural history, Academy of Geneva, 1816–35.

Carlyle, Jane Baillie Welsh (1801–66). Married Thomas Carlyle in 1826.

Carlyle, Thomas (1795–1881). Essayist and historian.

Cattell, John. Florist, nurseryman, and seedsman in Westerham, Kent.

Chambers, Robert (1802–71). Publisher, writer, and geologist. Anonymous author of *Vestiges of the natural history of creation* (1844).

Christopher, Robert Adam (1804–7). Of Bloxham Hall, Lincolnshire. MP for North Lincolnshire, 1837–57.

Clark, James (1788–1870). London physician. Physician in ordinary to Queen Victoria, 1837.

Clift, William (1775–1849). Naturalist. Curator of the Hunterian Museum, Royal College of Surgeons, 1793–1844.

Cook, James (1728–79). Commander of several voyages of discovery. Circumnavigated the world, 1768–71 and 1772–5.

Cooper, Antony Ashley, 7th earl of Shaftesbury (1801–85). Whig politician and philanthropist. Urged reform of laws protecting factory operatives, colliery workers, and chimney-sweeps.

Corfield, Richard Henry (1804–97). Attended Shrewsbury School, 1816–19. CD stayed with him in Valparaiso in 1834 and 1835.

Covington, Syms (1816?–61). Became CD's servant on the *Beagle* in 1833 and remained with him as assistant, secretary, and servant until 1839. Emigrated to Australia in 1839.

Cowper, William (1731–1800). Poet. Translated the works of Homer.

Crawfurd, John (1783–1863). Orientalist. Held several civil and political posts in Java, India, Siam, and Cochin China. Returned to England in 1827.

Cresy, Edward (1792–1858). Architect and civil engineer. A neighbour of CD's in Down, Kent.

Cuming, Hugh (1791–1865). Naturalist and traveller.

Cuvier, Georges (1769–1832). French systematist, comparative anatomist, palaeontologist, and administrator. Professor of natural history, Collège de France, 1800–32; professor of comparative anatomy, Muséum d'Histoire Naturelle, 1802–32.

Dana, James Dwight (1813–95). American geologist and zoologist. Naturalist with the United States Exploring Expedition to the Pacific, 1838–42. An editor of the *American Journal of Science and Arts* from 1846. Professor of geology, Yale, 1856–90.

Darwin, Anne Elizabeth (Annie) (1841–51). CD's oldest daughter.

Darwin, Caroline Sarah (1800–88). CD's sister. Married Josiah Wedgwood III in 1837.

Darwin, Catherine. *See* Darwin, Emily Catherine.

Darwin, Charles Waring (1856–8). Youngest child of CD. Died of scarlet fever.

Darwin, Elizabeth (Lizzie, Bessy) (1847–1926). CD's daughter.

Darwin, Emily Catherine (Catherine, Catty, Kitty) (1810–66). CD's sister. Married Charles Langton in 1863.

Darwin, Emma (1808–96). Youngest daughter of Josiah Wedgwood II. Married CD, her cousin, in 1839.

Darwin, Erasmus Alvey (Ras) (1804–81). CD's brother. Matriculated Christ's College, Cambridge, 1822. At Edinburgh University, 1825–6. Qualified but never practised as a physician. Lived in London from 1829 to his death.

Darwin, Francis (Frank) (1848–1925). CD's son. Botanist. BA, Trinity College, Cambridge, 1870.

Darwin, George Howard (1845–1912). CD's son. Mathematician. BA, Trinity College, Cambridge, 1868.

Darwin, Henrietta Emma (Etty) (1843–1927). CD's daughter. Assisted CD with some of his work.

Darwin, Horace (1851–1928). CD's son. Civil engineer. BA, Trinity College, Cambridge, 1874.

Darwin, Leonard (Lenny) (1850–1943). CD's son. Military officer and instructor. Attended the Royal Military Academy, Woolwich.

Darwin, Marianne (1798–1858). CD's oldest sister. Married Henry Parker in 1824.

Darwin, Mary Eleanor (September–October 1842). CD's third child.

Darwin, Robert Waring (1766–1848). CD's father. Physician. Had a large practice in Shrewsbury. Son of Erasmus Darwin and his first wife, Mary Howard. Married Susannah Wedgwood in 1796.

Darwin, Susan Elizabeth (Granny) (1803–66). CD's sister. Lived at her parental home, The Mount, Shrewsbury, until her death.

Darwin, Susannah. *See* Wedgwood, Susannah.

Darwin, William Erasmus (1839–1914). CD's oldest child. Attended Rugby School. BA, Christ's College, Cambridge, 1862. Banker in Southampton.

Davy, John (1790–1868). Brother of Humphry Davy. Served in the army medical service in Ceylon, the Mediterranean islands, and the West Indies.

De la Beche, Henry Thomas (1796–1855). Geologist. First director of the Geological Survey of Great Britain, 1835–55.

Denny, Henry (1803–71). Entomologist. Curator, museum of the Literary and Philosophical Society of Leeds.

Derby, Lord. *See* Stanley, Edward Smith.

Duncan, Andrew, the elder (1744–1828). Professor of physiology, Edinburgh University, 1790–1821.

Duncan, Andrew, the younger (1773–1832). Professor of materia medica, Edinburgh University, 1821–32.

Earle, Augustus (1793–1838). Artist and traveller. Joined the *Beagle* as artist in 1831. Left at Montevideo due to ill health; his position was taken by Conrad Martens.

Ehrenberg, Christian Gottfried (1795–1876). German naturalist, microscopist, and traveller. Studied the development of coral reefs. Professor at Berlin University.

Ellis, William (1794–1872). Missionary in South Africa and the South Sea Islands. Chief foreign secretary to the London Missionary Society.

Empson, William (1791–1852). Barrister. Professor of general polity and the laws of England at the East India Company College, Haileybury, 1824–52. Editor of the *Edinburgh Review*, 1847–52.

Endlicher, Stephan Ladislaus (1804–49). German botanist.

Eyton, Thomas Campbell (1809–80). Shropshire naturalist. Cambridge contemporary of CD's. Built a museum at his estate in Eyton, Shropshire, housing a fine collection of skins and skeletons of European birds.

Falconer, Hugh (1808–65). Palaeontologist and botanist. Superintendent of the botanic garden, Saharanpur, India, 1832–42. Superintendent of the Calcutta botanic garden and professor of botany, Calcutta Medical College, 1848–55.

Faraday, Michael (1791–1867). Assistant to Humphry Davy at the Royal Institution, 1812; director of the laboratory, 1825. Contributed extensively to the fields of electrochemistry, magnetism, and electricity.

FitzRoy, Robert (1805–65). Naval officer, hydrographer, and meteorologist. Commander of HMS *Beagle*, 1828–36. MP for Durham, 1841–3. Governor of New Zealand, 1843–5. Chief of the meteorological department, Board of Trade, 1854.

Forbes, Edward (1815–54). Zoologist, botanist, and palaeontologist. Professor of botany, King's College, London, and curator of the museum of the Geological Society, 1842. Palaeontologist with

Forbes, Edward, cont.
the Geological Survey, 1844–54. Professor of natural history, Edinburgh University, 1854.

Fox, Henry Stephen (1791–1846). Diplomat, serving in Brazil and, subsequently, Washington, DC.

Fox, William Darwin (1805–80). Clergyman. CD's second cousin. A close friend of CD's at Cambridge who shared his enthusiasm for entomology. Rector of Delamere, Cheshire, 1838–73.

Franklin, John (1786–1847). Naval officer and Arctic explorer. Lieutenant-governor of Van Diemen's Land (Tasmania), 1837–43. Leader of the 1845 expedition in search of a north-west passage during which all hands perished.

Fries, Elias Magnus (1794–1878). Swedish botanist. Professor of botany at Uppsala University, 1835.

Gärtner, Karl Friedrich von (1772–1850). German physician and botanist. Practised medicine in Calw, Germany, from 1802. Studied plant hybridisation.

Gay, Claude (1800–73). French naturalist and traveller. Professor of physics and chemistry at Santiago College (Chile), 1828–42.

Geoffroy Saint-Hilaire, Etienne (1772–1844). French zoologist. Professor of zoology, Muséum d'Histoire Naturelle, 1793.

Geoffroy Saint-Hilaire, Isidore (1805–61). French zoologist. Succeeded his father, Etienne, as professor at the Muséum d'Histoire Naturelle in 1841. Professor of zoology, Paris, 1850.

Gmelin, Johann Georg (1709–55). Naturalist and explorer. Professor of chemistry and natural history, Academy of Sciences, St Petersburg, 1731–47; of medicine, botany, and chemistry, Tübingen University, 1749.

Gosse, Philip Henry (1810–88). Naturalist, traveller, and writer. Studied marine invertebrates.

Gould, Augustus Addison (1805–66). American physician and conchologist. Practised medicine in Boston, Massachusetts.

Gould, John (1804–81). Ornithologist and artist. Taxidermist to the Zoological Society of London, 1826–81. Described the birds collected on the *Beagle* voyage.

Gray, Asa (1810–88). American botanist. Fisher Professor of natural history, Harvard University, 1842–73.

Gray, John Edward (1800–75). Naturalist. Assistant keeper of zoological collections at the British Museum, 1824; keeper, 1840–74.

Greville, Robert Kaye (1794–1866). Botanist. Made botanical tours in the Scottish highlands. MP for Edinburgh, 1856.

Grey, Charles, 2d Earl (1764–1845). Statesman. Prime minister, 1831–4.

Grey, George (1812–98). Army officer and Australian explorer. Governor of South Australia, 1841–5; of New Zealand, 1845–53 and 1861–8; of Cape Colony, 1854–61. Prime minister of New Zealand, 1877–9. Knighted, 1848.

Gully, James Manby (1808–83). Physician. Successful practitioner of the water treatment at his hydropathic establishment in Malvern, 1842–72.

Hawley, Richard Maddock. Physician.

Hearne, Samuel (1745–92). Explorer and colonial administrator in Canada.

Henslow, Frances Harriet (1825–74). Daughter of John Stevens Henslow. Married Joseph Dalton Hooker in 1851.

Henslow, John Stevens (1796–1861). Clergyman, botanist, and mineralogist. Professor of botany, Cambridge University, 1825–61. Rector of Hitcham, Suffolk, 1837–61. CD's teacher and friend.

Herbert, William (1778–1847). Naturalist, classical scholar, linguist, politician, and clergyman. Noted for his work on plant hybridisation. Rector of Spofforth, Yorkshire, 1814–40. Dean of Manchester, 1840–7.

Herschel, John Frederick William (1792–1871). Astronomer, mathematician, chemist, and philosopher. Master of the Royal Mint, 1850–5.

Higgins, John (1796–1872). Land agent. Agent for CD's farm at Beesby, Lincolnshire.

Holland, Henry (1788–1873). Physician. Distant relative of the Darwins and Wedgwoods. Physician in ordinary to Queen Victoria, 1852. President of the Royal Institution for many years.

Hooker, Frances Harriet. *See* Henslow, Frances Harriet.

Hooker, Joseph Dalton (1817–1911). Assistant director, Royal Botanic Gardens, Kew, 1855–65; director, 1865–85. Worked chiefly on taxonomy and plant geography. Son of William Jackson Hooker. Friend and confidant of CD.

Hooker, William Jackson (1785–1865). Botanist. Established the Royal Botanic Gardens at Kew in 1841 and served as first director. Father of Joseph Dalton Hooker.

Hope, Frederick William (1797–1862). Entomologist and clergyman. Gave his collection of insects to Oxford University and founded a professorship of zoology, 1849.

Hope, Thomas Charles (1766–1844). Professor of chemistry at Edinburgh University, 1799–1843.

Horner, Leonard (1785–1864). Geologist and educationist. A promoter of science-based education at all social levels. Father-in-law of Charles Lyell.

Humboldt, Friedrich Wilhelm Heinrich Alexander (Alexander) von (1769–1859). Eminent Prussian naturalist and traveller.

Huxley, Thomas Henry (1825–95). Assistant surgeon in HMS *Rattlesnake*, 1846–50, during which time he investigated marine invertebrates. Lecturer on natural history, Royal School of Mines, 1854; professor, 1857. Naturalist to the Geological Survey of Great Britain, 1855. Fullerian professor of physiology, Royal Institution, 1863–7. Hunterian professor of comparative anatomy, Royal College of Surgeons, 1863–9.

Impey. CD's college servant at Christ's College, Cambridge.

Jeffrey, Francis (1773–1850). Scottish judge, critic, and Whig politician. Editor of the *Edinburgh Review*, 1803–29.

Jenyns, Leonard (1800–93). Naturalist and clergyman. Brother-in-law of John Stevens Henslow. Vicar of Swaffham Bulbeck, Cambridgeshire, 1828–49. Described the *Beagle* fish specimens.

Johnson, Henry (1802/3–81). Physician. Contemporary of CD's at Shrewsbury School and Edinburgh University. Senior physician, Shropshire Infirmary.

Jussieu, Adrien Henri Laurent de (1797–1853). French botanist. Professor of botany, Muséum d'Histoire Naturelle, Paris, 1826.

Kennedy, Benjamin Hall (1804–89). Clergyman and teacher. Headmaster of Shrewsbury School, 1836–66. Regius professor of Greek, Cambridge University, 1867–89.

King, Philip Gidley (1817–1904). Eldest son of Phillip Parker King. Midshipman in the *Beagle*, 1831–6. Lived in Australia from 1836.

King, Phillip Parker (1793–1856). Naval officer and hydrographer. Commander of the *Adventure* and *Beagle* on the first surveying expedition to South America, 1826–30. Settled in Australia in 1834.

Kölreuter, Joseph Gottlieb (1733–1806). German botanist. Professor of natural history, Karlsruhe. Carried out extensive work on plant hybridisation.

Biographical Register

Lamarck, Jean Baptiste Pierre Antoine de Monet de (1744–1829). Professor of zoology, Muséum d'Histoire Naturelle, 1793. Believed in spontaneous generation and the progressive development of animal types; propounded a theory of transmutation.

Lamb, Henry William, 2d Viscount Melbourne (1779–1848). Statesman. Home Secretary under Lord Grey, 1830–4. Prime minister, 1835–41.

Lamouroux, Jean Vincent Félix (1776–1825). French naturalist. Professor of natural history, Caen, 1810.

Langton, Charles (1801–86). Rector of Onibury, Shropshire, 1832–40. Resided at Maer, Staffordshire, 1840–6; at Hartfield Grove, Sussex, 1847–62. Married Charlotte Wedgwood in 1832. After her death, married Emily Catherine Darwin in 1863.

Langton, Charlotte. See Wedgwood, Charlotte.

Leighton, Colonel. See Leighton, Francis Knyvett.

Leighton, Francis Knyvett (1772–1834). Army officer. A close friend of Robert Waring Darwin's.

Lewis, John. Carpenter in Down, Kent.

Lindley, John (1799–1865). Botanist and horticulturist. Professor of botany, London University (later University College London), 1828–60. Editor of the *Gardeners' Chronicle* from 1841.

Linnaeus (Carl von Linné) (1707–78). Swedish botanist and zoologist. Proposed a system for the classification of the natural world and reformed scientific nomenclature.

Lizars, John (1787?–1860). Surgeon and teacher of anatomy. Professor of surgery, Royal College of Surgeons, Edinburgh, 1831.

Loyd, Samuel Jones, 1st Baron Overstone (1796–1883). Banker. MP for Hythe, 1819–26.

Lubbock, John, 4th baronet and 1st Baron Avebury (1834–1913). Banker, politician, and naturalist. Son of John William Lubbock and a neighbour of CD's in Down. Studied entomology and anthropology. An active supporter of CD's theory of natural selection.

Lubbock, John William, 3d baronet (1803–65). Astronomer, mathematician, and banker. First vice-chancellor of London University, 1837–42. CD's neighbour in Down.

Lyell, Charles (1797–1875). Uniformitarian geologist whose *Principles of geology* (1830–3) and *Elements of geology* (1838) appeared in many editions. Travelled widely and published accounts of his trips to the United States. Scientific mentor and friend of CD.

Biographical Register

Macarthur, William (1800–82). Australian horticulturist, viniculturist, and amateur botanist. Member of the legislative council of New South Wales, 1849–55 and 1864–82.

MacCulloch, John (1773–1835). Physician, chemist, and geologist.

Mackintosh, Frances (Fanny) (1800–89). Married Hensleigh Wedgwood in 1832.

Macleay, William Sharp (1792–1865). Diplomat and naturalist. Stationed in Havana, 1825–37. Emigrated to Australia in 1839 and established a botanic garden at Elizabeth Bay, Sydney. Originator of the quinary system of taxonomy.

Mahon, Lord. *See* Stanhope, Philip Henry.

Malthus, Thomas (1766–1834). Clergyman and political economist. First professor of history and political economy at the East India Company College, Haileybury, 1805–34. Quantified the relationship between growth in population and food supplies in his *Essay on the principle of population* (1798).

Mantell, Walter Baldock Durrant (1820–95). Geologist, naturalist, and politician. Made important studies of the *Dinornis* beds of New Zealand. Commissioner of Crown Lands, Otago, in 1851. Elected for Wallace, 1861. Member of the Legislative Council, 1866–95.

Martens, Conrad (1801–78). Landscape painter. Joined HMS *Beagle* in Montevideo in 1833 and served as draughtsman until 1834. Settled in Australia in 1835.

Martineau, Harriet (1802–76). Author, reformer, and traveller.

Maury, Matthew Fontaine (1806–73). American naval officer, hydrographer, and meteorologist. Head of the National Observatory, 1844–61.

Melbourne, Viscount. *See* Lamb, Henry William.

Milne-Edwards, Henri (1800–85). French zoologist. Professor of hygiene and natural history, Ecole Centrale des Arts et Manufactures, 1832. Professor of entomology, Muséum d'Histoire Naturelle, 1841; professor of mammalogy, 1861.

Mirbel, Charles François Brisseau de (1776–1854). French botanist. Professor-administrator of the Jardin des Plantes, 1829.

Mitford, William (1744–1827). Historian and politician.

Monro, Alexander, tertius (1773–1859). Anatomist. Professor of medicine, surgery, and anatomy, Edinburgh University, 1817–46.

Müller, Johannes Peter (1801–58). German comparative anatomist, physiologist, and zoologist. Professor of anatomy and physiology, Berlin University, 1833.

Murray, John (1808–92). CD's publisher from 1845.

Narbrough, John (1640–88). Admiral. Commissioner of the navy, 1680–7.

Orbigny, Alcide Charles Victor Dessalines d' (1802–57). French palaeontologist. Professor of palaeontology, Muséum d'Histoire Naturelle, 1853.

Overstone, Lord. *See* Loyd, Samuel Jones.

Owen, Fanny (Frances) Mostyn. Second daughter of William Mostyn Owen Sr of Woodhouse. Married Robert Myddelton Biddulph in 1832. A close friend and neighbour of CD's before the *Beagle* voyage.

Owen, Richard (1804–92). Anatomist. Hunterian professor, Royal College of Surgeons, 1836–56. Superintendent of the natural history departments, British Museum, 1856–84. Described the *Beagle* fossil mammal specimens.

Owen, William Mostyn, Sr. Lieutenant, Royal Dragoons. Squire of Woodhouse, Shropshire.

Paley, William (1743–1805). Anglican clergyman and philosopher who propounded a popular system of natural theology.

Parker, Henry (1788–1856). Physician to the Shropshire Infirmary. Married Marianne Darwin in 1824.

Peacock, George (1791–1858). Tutor in mathematics at Trinity College, Cambridge, 1823–39. Lowndean professor of geometry and astronomy, Cambridge University, 1837. Dean of Ely, 1839–58.

Phillips, John (1800–74). Geologist. Professor of geology, King's College, London, 1834–40. Palaeontologist to the Geological Survey of Great Britain, 1840–4. Deputy reader in geology, Oxford University, 1853; professor, 1860–74.

Price, John (1803–87). Welsh scholar, naturalist, and schoolmaster. Assistant master at Shrewsbury School, 1826–7. Private tutor in Chester.

Pulleine, Robert (1806–68). Rector of Kirkby-Wiske, 1845–68.

Quetelet, Lambert Adolphe Jacques (1796–1874). Belgian statistician. Astronomer, Brussels Royal Observatory, 1828–74. Secretary, Académie Royale des Sciences et Belles-Lettres, Brussels, 1834–74.

Raja of Sarawak. *See* Brooke, James.

Ramsay, Andrew Crombie (1814–91). Geologist, Geological Survey of Great Britain, 1841; senior director for England and Wales, 1862; director-general, 1871–81. Professor of geology, University College London, 1847–52.

Ramsay, Marmaduke (d. 1831). Fellow and tutor, Jesus College, Cambridge, 1819–31.

Richardson, John (1787–1865). Arctic explorer and naturalist. Surgeon and naturalist on John Franklin's polar expeditions, 1819–22 and 1825–7. Surgeon to Chatham division of marines, 1824–38. Physician to the Royal Hospital, Haslar, in 1838. Conducted search expedition for John Franklin, 1847–9.

Rogers, Henry (1806?–77). Congregationalist minister. Professor of English language and literature, University College London, 1837; of English, mathematics, and mental philosophy, Spring Hill College, Birmingham, 1839.

Ross, James Clark (1800–62). Naval officer and polar explorer. Discovered the north magnetic pole in 1831. Commander of an expedition to the Antarctic, 1839–43; of a search expedition for John Franklin, 1848–9.

Roxburgh, William (1751–1815). Botanist and surgeon. Superintendent of the Calcutta Botanic Garden, 1793.

Royle, John Forbes (1799–1858). Surgeon and naturalist in the service of the East India Company. Superintendent of the botanic garden in Saharanpur, India, 1823–31. Professor of materia medica, King's College, London, 1837.

Sabine, Edward (1788–1883). Geophysicist and army officer. General secretary, British Association for the Advancement of Science, 1838–59. Foreign secretary, Royal Society, 1845–50; treasurer, 1850–61; president, 1861–71.

Saunders, William Wilson (1809–79). Underwriter at Lloyd's. President of the Entomological Society, 1841–2 and 1856–7. Treasurer of the Linnean Society, 1861–73.

Say, Thomas (1787–1834). American entomologist and conchologist. Curator of the American Philosophical Society, 1821–7; professor of natural history, University of Pennsylvania, 1822–8.

Scherzer, Karl von (1821–1903). Viennese scientific traveller and diplomat. Austrian consul in London, 1875–8.

Schleiden, Matthias Jacob (1804–81). German botanist and naturalist. Recognised as a founder of cell theory.

Sedgwick, Adam (1785–1873). Geologist and clergyman. Woodwardian professor of geology, Cambridge University, 1818–73. Canon of Norwich, 1834–73.

Seymour, Edward Adolphus, 11th duke of Somerset (1775–1855). President of the Linnean Society, 1834–7.

Shaftesbury, Lord. *See* Cooper, Antony Ashley.

Silliman, Benjamin (1779–1864). American chemist, geologist, and mineralogist. Professor of chemistry and natural history, Yale University, 1802–53. Founder and first editor of the *American Journal of Science and Arts*, 1818.

Smith, Andrew (1797–1872). Army surgeon stationed in South Africa, 1821–37. An authority on South African zoology. Principal medical officer at Fort Pitt, Chatham, 1837; deputy inspector-general, 1845. Director-general, Army Medical Department, 1853–8.

Smith, Charles Hamilton (1776–1859). Army officer and writer on natural history. FRS 1824.

Smith, Frederick (1805–79). Entomologist in the zoological department of the British Museum from 1849.

Smith, James (1782–1867). Known as Smith of Jordanhill. Scottish antiquarian, numismatist, and geologist. Partner in a firm of West India merchants.

Somerset, Duke of. *See* Seymour, Edward Adolphus.

Spencer, Herbert (1820–1903). Writer. Civil engineer on the railways, 1837–41 and 1844–6. Sub-editor of the *Economist*, 1848–53. Author of papers on evolution and numerous works on philosophy and the social sciences.

Stanhope, Philip Henry, 4th Earl (1781–1855). MP for Wendover, 1806–7; for Kingston-upon-Hull, 1807–12; for Midhurst, 1812–16.

Stanhope, Philip Henry, 5th Earl (1805–75). Historian. Known as Viscount Mahon from 1816 until 1855 when he succeeded to the earldom. MP for Wootton Bassett, 1830–2; for Hertford, 1832–3 and 1835–52. President of the Society of Arts, 1846–75. Resided at the family seat in Chevening, Kent.

Stanley, Edward Smith, 13th earl of Derby (1775–1851). MP for Preston, 1796–1812; for Lancashire, 1812–32. President of the Zoological Society.

Stephens, Catherine (1794–1882). English soprano and actress.

Stephens, James Francis (1792–1852). Entomologist and zoologist. Employed in the Admiralty office, Somerset House, 1807–45. Assisted in arranging the insect collection at the British Museum.

Stevens, John Crace. Auctioneer and natural history dealer in Covent Garden.

Stokes, John Lort (1812–85). Naval officer. Served in HMS *Beagle* as midshipman, 1826–31; as mate and assistant surveyor, 1831–7; as lieutenant, 1837–41; as commander, 1841–3.

Strickland, Hugh Edwin (1811–53). Geologist and zoologist. An advocate of reform in zoological nomenclature.

Strzelecki, Paul Edmond de (1797–1873). Polish-born explorer and geologist. Explored the Australian interior, 1839–40. Naturalised as a British subject in 1845.

Stutchbury, Samuel (1798–1859). Geologist and naturalist. Curator of the museum of the Bristol Philosophical Institution, 1831. Surveyor in Australia, 1850–5.

Sulivan, Bartholomew James (1810–90). Naval officer and hydrographer. Lieutenant in HMS *Beagle*, 1831–6.

Sutherland-Leveson-Gower, George Granville, 2d duke of Sutherland (1786–1861). MP for St Mawes, Cornwall, 1808–12; for Newcastle-under-Lyme, 1812–15; for Staffordshire, 1815–20. Lord lieutenant of Sutherland, 1831–61; of Shropshire, 1839–45.

Syme, Patrick (1774–1845). Flower painter and drawing-master.

Tegetmeier, William Bernhard (1816–1912). Editor, journalist, lecturer, and naturalist. Pigeon fancier and expert on fowls and bees.

Thorley, Catherine A. Governess at Down House, 1850–6. Present at Anne Elizabeth Darwin's death in Malvern in 1851.

Wallace, Alfred Russel (1823–1913). Collector in the Amazon, 1848–52; in the Malay Archipelago, 1854–62. Independently formulated a theory of natural selection in 1858.

Waterhouse, George Robert (1810–88). Naturalist. Joined natural history department of the British Museum, 1843; keeper, geology department, 1857–80. Described mammalian and entomological specimens from the *Beagle* voyage.

Watson, Hewett Cottrell (1804–81). Botanist, phytogeographer, and phrenologist. Published various guides to the distribution of British plants.

Wedgwood, Charlotte (1797–1862). Emma Darwin's sister. Married Charles Langton in 1832.

Wedgwood, Elizabeth. *See* Allen, Elizabeth.

Wedgwood, Emma. *See* Darwin, Emma.

Wedgwood, Frances Mackintosh. *See* Mackintosh, Frances.

Wedgwood, Henry Allen (1799–1885). Barrister. Married his cousin, Jessie Wedgwood in 1830.

Wedgwood, Hensleigh (1803–91). Emma Darwin's brother. Philologist and barrister. Metropolitan police magistrate, Lambeth, 1832–7.

Wedgwood, Jessie (1804–72). CD's and Emma Darwin's cousin. Married Henry Allen Wedgwood in 1830.

Wedgwood, Josiah, I (1730–95). Master-potter. Founded the Wedgwood pottery works at Etruria, Staffordsire. Grandfather of CD and Emma Darwin.

Wedgwood, Josiah, II (Uncle Jos) (1769–1843). Of Maer Hall, Staffordshire. Master-potter of Etruria. Whig MP for Stoke-on-Trent, 1832–4. Emma Darwin's father.

Wedgwood, Josiah, III (1795–1880). Partner in the Wedgwood pottery in Staffordshire until 1841, when he moved to Leith Hill Place, Surrey. Emma Darwin's brother. Married Caroline Sarah Darwin in 1837.

Wedgwood, Mary Anne (1796–8). Emma Darwin's sister.

Wedgwood, Sarah Elizabeth (Elizabeth, Aunt Sarah) (1793–1880). Emma Darwin's sister. Resided at Maer Hall, Staffordshire, until 1847, then at the Ridge, Hartfield, Sussex, 1847–62.

Wedgwood, Susannah (1765–1817). CD's mother. Daughter of Josiah Wedgwood I. Married Robert Waring Darwin in 1796.

Wharton, Henry James (1798–1859). Vicar of Mitcham, Surrey, 1846–59. Tutor to William Erasmus Darwin, 1850–1.

Whately, Thomas (d. 1772). Politician and man of letters.

Whewell, William (1794–1866). Mathematician and historian and philosopher of science. Master of Trinity College, Cambridge, 1841–66. Professor of moral philosophy, Cambridge, 1838–55.

Whitley, Charles Thomas (1808–95). Attended Shrewsbury School, 1821–6. Reader in natural philosophy and mathematics, Durham University, 1833–55. Vicar of Bedlington, Northumberland, 1854–95.

Wickham, John Clements (1798–1864). Naval officer and magistrate. First-lieutenant in the *Beagle*, 1831–6; commander, 1837–41, surveying the Australian coast.

Williams, John (1796–1839). Missionary in the Pacific. Killed and eaten by natives of Eromanga Island.

Wollaston, Thomas Vernon (1822–78). Entomologist and conchologist. Passed many winters in Madeira where he collected insects and shells.

Wood, Alexander Charles (b. 1810). A colonial land and emigration commissioner. Robert FitzRoy's cousin.

Woodward, Samuel Pickworth (1821–65). Naturalist. Sub-curator, Geological Society, 1839–45. Professor of geology and natural history, Royal Agricultural College, Cirencester, 1845. Assistant, department of geology and mineralogy, British Museum, 1848–65.

Worsley, Charles Anderson, 2d earl of Yarborough (1809–62). MP for Lincolnshire, 1831–2; for North Lincolnshire, 1832–46.

Yarborough, Lord. *See* Worsley, Charles Anderson.

Bibliography of Biographical Sources

J. E. Auden, ed. *Shrewsbury School register 1734–1908*. Oswestry: Woodall, Minshall, Thomas and Co., Caxton Press, 1909.

J. Balteau *et al. Dictionnaire de biographie Française*. 17 vols. and 4 fascicles of vol. 18. (A–Jumelle). Paris: Libraire Letouzey et Ané, 1933–92.

Frederic Boase. *Modern English biography containing many thousand concise memoirs of persons who have died since the year 1850*. 3 vols. and supplement (3 vols.). Truro: printed for the author, 1892–1921.

H. F. Burke. *Pedigree of the family of Darwin*. Privately printed, 1888. [Reprinted in facsimile in R. B. Freeman, *Darwin pedigrees*. London: printed for the author, 1984.]

John Burke. *Burke's genealogical and heraldic history of the landed gentry*. 1–18 editions. London: Burke's Peerage Ltd., 1836–1972.

Ray Desmond. *Dictionary of British and Irish botanists and horticulturists including plant collectors, flower painters and garden designers*. New edition, revised and completely updated. London: Taylor & Francis and the Natural History Museum. Bristol, Pa.: Taylor & Francis, 1994.

R. B. Freeman. *Charles Darwin: a companion*. Folkestone, Kent: William Dawson and Sons. Hamden, Conn.: Archon books, Shoe String Press, 1978.

Pamela Gilbert. *A compendium of the biographical literature on deceased entomologists*. London: British Museum (Natural History), 1977.

C. C. Gillispie, ed. *Dictionary of scientific biography*. 14 vols., supplement, and index. New York: Charles Scribner's Sons, 1970–80.

George Grove. *The new Grove dictionary of music and musicians*. Edited by Stanley Sadie. 20 vols. London: Macmillan, 1980.

Leonard Huxley ed. *Life and letters of Sir Joseph Dalton Hooker*. 2 vols. London: John Murray, 1918.

Henrietta Litchfield, ed. *Emma Darwin: a century of family letters 1792–1896*. 2 vols. London: John Murray, 1915.

Neue deutsche Biographie. Under the auspices of the Historical Commission of the Bavarian Academy of Sciences. 17 vols. (A–Moller). Berlin: Duncker and Humblot, 1953–94.

Douglas Pike and Bede Nairn, eds. *Australian dictionary of biography: 1788–1850; 1851–1890.* 6 vols. Melbourne: Melbourne University Press, 1966–76.

Post Office directory of the six home counties, viz., Essex, Herts, Kent, Middlesex, Surrey and Sussex. London, 1845–.

Post Office London directory. London, 1802–.

The provincial medical directory. London, 1847.

E. W. Richardson. *A veteran naturalist; being the life and work of W. B. Tegetmeier.* London: Witherby & Co, 1916.

W. A. S. Sarjeant. *Geologists and the history of geology: an international bibliography from the origins to 1978.* 5 vols. and 2 supplements. London: Macmillan, 1980–7.

Leslie Stephen and Sidney Lee, eds. *Dictionary of national biography.* 63 vols. and 2 supplements (6 vols.). London: Smith, Elder, and Co., 1885–1912. H. W. C. Davis *et al.*, eds. *Dictionary of national biography 1912–85.* 8 vols. London: Oxford University Press, 1927–90.

J. A. Venn. *Alumni Cantabrigienses. A biographical list of all known students, graduates and holders of office at the University of Cambridge, from the earliest times to 1900. Part II. From 1752 to 1900.* 6 vols. Cambridge: Cambridge University Press, 1940–54.

Constant von Wurzbach. *Biographisches Lexicon des Kaiserthums Oesterreich, enthaltend die Lebensskizzen der denkwürdigen Personen, welche 1750 bis 1850 im Kaiserstaate und in seinen Kronländern gelebt haben.* 60 vols. Vienna, 1856–90.

Bibliographical Note

The following bibliography contains a list of Darwin's works cited in the letters and notes. An asterisk indicates that a book is available in paperback.

Autobiography *: Nora Barlow, ed. *The autobiography of Charles Darwin 1809–1882. With original omissions restored.* London: Collins, 1958. Reprinted New York: Norton, 1993.

Beagle diary: R. D. Keynes, ed. *Charles Darwin's Beagle diary.* Cambridge: Cambridge University Press, 1988.

Collected papers *: P. H. Barrett, ed. *The collected papers of Charles Darwin.* 2 vols. Chicago and London: University of Chicago Press, 1977.

Correspondence: *The correspondence of Charles Darwin.* Vols. 1–9. Cambridge: Cambridge University Press, 1985–95.

Coral reefs *: Charles Darwin. *The structure and distribution of coral reefs.* Berkeley: University of Arizona Press, 1984.

Journal: Gavin de Beer, ed. 'Darwin's Journal'. *Bulletin of the British Museum (Natural History Historical Series)* 2 (pt 1): 4–21. London, 1959.

Journal of researches. See *Voyage of the Beagle.*

Living Cirripedia: Charles Darwin. [Vol. 1] *A monograph on the sub-class Cirripedia ... The Lepadidæ; or, pedunculated cirripedes.* [Vol. 2] *... The Balanidæ, (or sessile cirripedes); the Verrucidæ, etc.* London: The Ray Society, 1851 and 1854.

Natural selection *: R. C. Stauffer, ed. *Charles Darwin's Natural Selection, being the second part of his big species book written from 1856 to 1858.* Cambridge: Cambridge University Press, 1975.

Notebooks: P. H. Barrett *et al.*, eds. *Charles Darwin's notebooks, 1836–1844.* Cambridge: Cambridge University Press for the British Museum (Natural History), 1987.

Origin *: Charles Darwin. *On the origin of species by means of natural selection, or the preservation of favoured races in the struggle for life.* London: John Murray, 1859. (*The origin of species* is available in several paperback editions. A facsimile of the first edition with an introduction by Ernst Mayr, published by Harvard University Press, 1966, has been used in this volume.)

South America: Charles Darwin. *Geological observations on South America*. London: John Murray, 1846.

Voyage of the Beagle: Charles Darwin. Originally published as the third volume of *Narrative of the surveying voyages of his Majesty's Ships Adventure and Beagle*, edited by Captain Robert FitzRoy with the subsidiary title, *Journal and remarks*, 1839. Published separately with several different titles, it is now generally known as *The voyage of the Beagle*. The edition cited in this volume is Janet Browne and Michael Neve, eds. *Charles Darwin, Voyage of the Beagle*.* London: Penguin, 1989.

Zoology: *The zoology of the voyage of H.M.S. Beagle* ... Edited and superintended by Charles Darwin. 5 vols. London: Smith Elder and Co., 1838–43.

Further Reading

Mea Allan. *Darwin and his flowers. The key to natural selection*. London: Faber & Faber. New York: Taplinger, 1977.

John Bowlby. *Charles Darwin. A biography*. London: Hutchinson, 1990.

P. J. Bowler. *Charles Darwin: the man and his influence*. Oxford: Blackwell, 1990.

P. J. Bowler. *Evolution: the history of an idea*. Revised ed. Berkeley and Los Angeles: University of California Press, 1989.

Janet Browne. *Charles Darwin. Voyaging*. New York and London: Knopf, 1995.

Francis Darwin, ed. *The life and letters of Charles Darwin, including an autobiographical chapter*. 3 vols. London: John Murray, 1887.

Adrian Desmond and J. R. Moore. *Darwin*. London: Michael Joseph, 1991.

S. J. Gould. *Ever since Darwin: reflections in natural history*. New York and London: Norton, 1974.

John Maynard Smith. *The theory of evolution*. Cambridge: Cambridge University Press, 1993.

Jonathan Miller and Borin Van Loon. *Darwin for beginners*. London: Writers and Readers, 1982.

Alan Moorehead. *Darwin and the Beagle*. Revised ed. London: Penguin, 1979.

D. M. Porter and P. W. Graham, eds. *The portable Darwin*. New York and London: Penguin, 1993.

Index

aberrant genera. *See* anomalous genera

Abrolhos Islands, 20; specimens from, 23

acquired instincts, 56, 213

Agassiz, Jean Louis Rodolphe, 97; on human races, 115; on embryo development, 132; expected to attack species theory, 161; and glaciation controversy, 180

agouti, 29

ammonite, 31

amphibians, South American, 25

analogies, 208

Anatifa spp., 112

Andes: CD's excursions to, 36, 42–3; geology, 43, 46–7; 'red snow' in, 44

anomalous/aberrant genera, 133–4

artificial selection, 178, 197

Ascidia spp., 127, 216

Athenæum Club, 66; Willy too young for membership, 151; Huxley not proposed for membership, 152–3, 155

Atlantic continent, 135, 155, 156, 156–8, 171

Australia: *Beagle* visit, 49, 50; geology of, 112; CD considers prospects in, 118

Austrian circumnavigation expedition, 165–6, 217

awards to scientists, 175–6

Babbage, Charles, invites CD to parties, 54

Bahia Blanca: CD writes from, 16, 52; *Beagle* calls at, 18, 51, 52; beauty of, 19; specimens from, 26, 59

Baily, John, 150

Baily, William Hellier, 132

barnacles, xxii, 121n.; nomenclature,

103, 111; CD's requests for specimens, 104, 107; CD acknowledges specimens, 117, 118; fossil, 162; hybrids, how produced, 182–3

Basingwaithe (ship), 36

Beagle voyage: CD offered position on, 11–15; begins from Plymouth, 15n., 16; to Bahia Blanca, 19, 51, 52; to Rio de Janeiro, 21; to Montevideo, 24; to Tierra del Fuego, 27, 29; to Falkland Islands, 28, 29, 32; to Santa Cruz, 34; to Chiloé, 35; to Valparaiso, 37; to Cape Tres Montes, 40, 41; to Chonos archipelago, 40; to Lima, 46; to Galápagos, 48; to New Zealand, 48, 49; to Tahiti, 48–9; to Australia, 49, 50; to Mauritius, 50; ends at Falmouth, 53

Beaufort, Francis: plans *Beagle*'s journey, 14, 15; FitzRoy requests CD's appointment confirmed, 15n.; CD seeks support for publications grant, 57–8

Bechstein, Johann Matthäus, 159, 216

bees: and flower fertilisation, 186–7, 191–2; collection of honeycombs, 197–8

Bell, Thomas, contributor, *Zoology*, 60n.

Bentham, George, 134; on species relatedness, 145

Berkeley, Miles Joseph, on seeds' resistance to salt water, 146

Bible, CD enjoys Gospels, 4

Biddulph, Robert Myddelton, 212

Binney, Edward William, pamphlet, 92, 215

Birch, Samuel, CD's inquiry on Chinese agriculture, 149

birds: of South America, 25; of Galápagos, 59; whether feet are dirty, 159;

Index

Darwin, Emma, cont.
confinements, 88, 114–15, 121n., 164, 214, 215, 217; reads Lyell's book on America, 108; plans visit to British Association, 109; grief after baby's death, 190

Darwin, Erasmus, on acquired instincts, 56, 213

Darwin, Erasmus Alvey: studies medicine at Edinburgh, 1–5; studies medicine in London, 5n.; visits Cambridge, 45; preparations for CD's return, 51; friendship with Harriet Martineau, 64; visits Carlyles with CD, 66–7; house-hunting with CD, 67–8; in poor health, 165

Darwin, Francis, birth, 215

Darwin, George Howard: birth, 88, 214; drawing all day, 122, 123; dislikes reading, 123–4; home-sick at school, 167

Darwin, Henrietta Emma, 87, 88, 214; recovering from illness, 190, 217

Darwin, Horace, birth, 121n.

Darwin, Leonard, 123; birth, 114–15, 215; letter from, 150

Darwin, Robert Waring: financial support for CD, 4, 41–2, 45, 51; advises CD against joining *Beagle*, 11, 12; CD seeks information on contagious diseases, 56–7; delight at CD's engagement, 65; advises on baby care, 71; ill at Shrewsbury, 102; death, 105, 107, 215; quoted, on best time to seek advice, 154

Darwin, Susan Elizabeth, 22; CD asks for letters, 3; CD welcomes letters from, 3, 52; letters to CD missing, 42; visits CD, 72; and child chimney-sweeps, 125

Darwin, William Erasmus: birth, xxii, 214; childhood, 71, 72, 87, 88; CD's plans for education, 115, 116; collects Lepidoptera, 115; learning to ride, 116–17; at Rugby, 117n., 123, 124, 150, 166–7; CD urges more pleasant manners, 122; to go to Cambridge,

187; studying at Cambridge, 194–5

Davy, John, on transport of fishes' ova, 136–7

De la Beche, Henry Thomas, CD's enquiry on Jamaican livestock, 73

Denny, Henry, CD seeks information on parasites, 85–6

Derby, Lord. *See* Stanley, Edward Smith

Dicentra, 191

dispersal. *See* species dispersal

divergence, 188

donkeys, shoulder-stripes on, 194

Down House, 74–6; beech tree felled, 150; 'weed garden', 170, 174–5

Duke of York (ship), 26

Duncan, Andrew (the elder), 3–4

Duncan, Andrew (the younger), 2

Earle, Augustus, 49, 212

earthquakes in Chile, 42, 44

Edinburgh: CD a student in, 1–5, 103; CD plans to visit, 8

Edwardsia, 186

Ellis, William, *Polynesian researches*, 56, 213

entomology: CD asks for identification of specimens, 6–7; CD plans excursions for, 8–9; CD asks for advice on mounting Diptera, 10; South American insect specimens, 23; excursions remembered, 59, 164; insects' role in pollination, 88; Wollaston's book on, 135

erratic boulders, 95, 132–3

Evans (Australian bookseller), 77

Evans (butler), 108

Evans, Mrs, 77, 108

extinction of species, 59, 134, 213

Eyton, Thomas Campbell, 10; CD advises South American visit, 20; marriage, 45; CD asks for help with work, 148, 159–60; collection of skeletons, 193

Falconer, Hugh: quarrel with Huxley, 155, 156; criticism of CD, 177, 195; CD seeks advice on anatomy, 183;

Index

Falconer, Hugh, cont.
on age of human species, 195
Falkland Islands, 27, 28, 29, 32
feral animals, 73
Fernando Norunho, 18, 19
fishes: possible mode of dispersal, 136–7; refuse to swallow seeds, 139; whether seed-eating, 159–60
FitzRoy, Robert, xxi, 1n.; encourages CD to join *Beagle*, 13–15; requests CD's appointment confirmed, 15n.; commissions schooners for surveying expedition, 24; purchases schooner, 29; schooner sold, 38, 40; resigns fearing insanity, 40; withdraws resignation and recovers, 40, 41; account of *Beagle* voyage, 51, 55, 63; changes route of homeward voyage, 52; paper defending missionaries, 54, 213; CD visits, 63; appointed Governor of New Zealand, 77; new command, 107
Flustra, 5n., 26, 31
Forbes, Edward, 83, 89; visits Down House, 102; on molluscs, 131; theory of Atlantic continent, 135, 155, 156, 170, 171, 172, 184
fossils: South American finds, 25, 30–1, 36, 43, 59; Cirripedia, 114, 161–2; human remains, 195
Fox, Henry Stephen, corresponds with CD, 39
Fox, William Darwin: CD seeks advice on entomology, 6–7; anxious about examinations, 8; visit to CD planned, 8–9; marriage, 45; entomological excursions with CD, 59, 164; CD seeks information on hybrids, 72–3; tenth child born, 124
Freischütz, Der (C. M. von Weber), 2, 211
Fries, Elias Magnus, on relatedness of species, 145
Fumaria, 191

Galápagos Islands, 46, 48, 165; birds and reptiles of, 59; plants of, 78, 88; organisms of, key to species theory, 80, 203; birds and sea-shells of, 81;

geology of, 112
Gardeners' Chronicle: Hooker's Indian discoveries described, 110; CD on seeds' resistance to sea-water, 137, 143, 146–7, 216; CD on fertilisation in Leguminosae, 195
Gärtner, Karl Friedrich von, data on plant hybridisation, 173
Gay, Claude, 36, 212
Geoffroy Saint-Hilaire, Isidore, 'loi du balancement', 91
Geological Society: CD elected, 54n.; CD reads papers, 58, 167; CD declines nomination as secretary, 60–2, 214; CD elected secretary, 62n.
glacial deposits, 132–3
glaciation, 180–1
Gosse, Philip Henry, CD's inquiry on molluscs, 171–2
Gould, Augustus Addison, on distribution of land-shells, 184
Gould, John: names new ostrich species, 56; contributor, *Zoology*, 60n.
Granby (T. H. Lister), 4, 211
grasses, 142, 144
Gray, Asa: CD seeks botanical information, 138–9, 168; manual of American flora, 142–4; sends list of 'close species', 144–5; sketch of CD's species theory sent to, 177–80, 190, 192; on hypothetical nature of CD's work, 182; on difficulty of defining species, 183
Gray, John Edward, 126; CD asks for help from, 96–7; CD seeks information on Chinese agriculture from, 149
Grey, Charles, 2d earl, resignation, 39, 212
Grey, George, offers help with scientific studies, 94–5
Gully, John Manby: treats CD hydropathically, 105–6, 108, 111; CD considers credulous, 116; treats Annie in last illness, 119–21; treats W. D. Fox, 167

hawks, pellets, 159

Index

Impey (college servant), 194, 198
India, Hooker's visit to, 100n., 110
intermediate forms, 101
Iquique, 47
island species, 161, 165, 173, 184, 204;
genus/species numbers, 80; plants
with hooked seeds, 80; distinctive
character of, 81, 88; whether insects
absent, 88; plants, 100–1, 113; CD's
views different from Wallace's, 185,
204
isolation, and numbers of species, 84

jaguars, mating of like-coloured, 173
Jamaica, animals of, 73
Jenyns, Leonard, 9–10, 33, 214; con-
tributor, *Zoology*, 60n.; sent copy of
Origin, 206
John Bull, 5, 211
Johnson, Henry: lodging very cheaply,
2–3; experiments with nitric oxide, 4
Journal, xxi, 58, 59n., 64n., 68, 201;
notes made on *Beagle* voyage, 19, 33;
preparation for press, 58, 68; second
edition published, 107; copy given to
Tegetmeier, 146; CD recommends
to Austrian explorers, 165
Jussieu, Adrien Henri Laurent de, 131

Keeling Islands, 50, 88
Kennedy, Benjamin Hall, 8
Kerguelen Land, 80, 88, 165
King, Philip Gidley, 18
King, Philip Parker: ammonite find by,
31; journal of, 63; CD sends kind
remembrances to, 108
Knowles, Sir F., 70
Kölreuter, Joseph Gottlieb, data on
plant hybridisation, 173

Lamarck, Jean Baptiste Pierre Antoine
de Monet de, 89–90
Lamoureux, Jean Vincent Félix, 212;
on corallines, 35
Langton, Charlotte (née Wedgwood),
22, 165
Leguminosae, 134; fertilisation of, 186–
7, 195

Leighton, Francis Knyvett, death, 44
leopards, mating of like-coloured, 173
Lepidodendron, 93, 94, 133
Lewis, John, 87
Lewis, William, 87
Lima, CD writes from, 46
Lindley, John, 129, 162
Linnæus: on principles of classification,
76; on survival of seeds at sea, 147
Linnean Society, xxiv, 192, 193, 194,
211, 217, 218
Living Cirripedia, xxii; preparation for
publication, 104, 107; complimen-
tary copies sent, 125n., 131–2; possible review by Huxley, 127–8
lizards, 140, 167
Lizars, John, 2
Lobelia fulgens, fertilisation, 183
Loyd, Samuel Jones, 1st Baron Overstone, 151
Lubbock, John, 103; living in Downe,
76; suggests numerical analysis of
data, 176–7; conversion to species
theory, 203
Lyell, Charles, 34, 47, 59, 82, 83;
on coral reefs, xxi; CD dines
with, 54n., 66; on CD's article
and fossil collection, 54, 213; CD
values help and friendship of, 58;
CD acknowledges debt to, 83;
Principles of geology, 83, 214; lends
cirripede collection to CD, 98;
visits Down House, 102, 152n.;
book on American visit, 108, 109,
215; encourages CD to approach
Mantell, 133; urges CD to publish
work on species, xxiv, 152, 153,
154; reproved by CD for disciples'
theories, 155, 156, 157; owns cat's
skeleton, 160; paper on Madeira,
171; on struggle for existence, 178;
on glaciation, 180–1; consulted by
CD on Wallace's paper, 184, 188,
189–90, 192; on theory of ocean
islands, 184; changing views on
species theory, 198, 205, 218; CD
asks advice on publication of *Origin*,
199–201

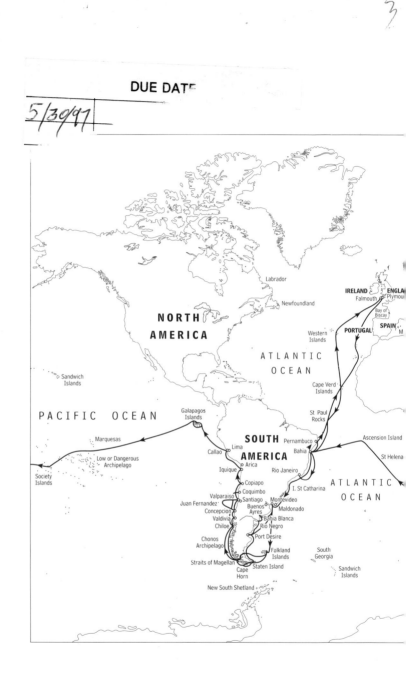